Land Warfare: Brassey's New Battlefield
Weapons Syst   es

                                                2

Command and
Control
Support Systems in the Gulf War

LAND WARFARE: Brassey's New Battlefield Weapons Systems
and Technology Series

Executive Editor: Colonel R G Lee OBE, Former Military Director of
Studies, Royal Military College of Science, Shrivenham,
UK.

Editor-in-Chief: Professor Frank Hartley, Vice Chancellor, Cranfield
Institute of Technology, UK.

The success of the first series on Battlefield Weapons Systems and
Technology and the pace of advances in military technology have prompted
Brassey's to produce a new Land Warfare series. This series updates subjects
covered in the original series and also covers completely new areas. The
books are written for military personnel who wish to advance their professional knowledge. In addition, they are intended to aid anyone who is inter-
ested in the design, development and production of military equipment.

Command and Control

Support Systems in the Gulf War

An account of the command and control information systems support to the British Army contribution to the Gulf War

M A Rice & A J Sammes

BRASSEY'S

LONDON * NEW YORK

Copyright © 1994 Brassey's (UK) Ltd

First English edition 1994

UK editorial offices:
Brassey's, 33 John Street, London WC1N 2AT

Orders:
Marston Book Services, PO Box 87, Oxford OX2 0DT

USA orders: Macmillan Publishing Company, Front and Brown Streets, Riverside, NJ 08075

Distributed in North America to booksellers and wholesalers by the Macmillan Publishing Company, NY 10022

Library of Congress Cataloging in Publication Data
Available

British Library Cataloging in Publication Data
A catalogue record for this book is available from the British Library

1 85753 015 2 Flexicover

M A Rice & A J Sammes have asserted their moral rights to be identified as authors of this work

Typeset by Florencetype Ltd, Kewstoke, Avon
Printed and bound in Great Britain by
Butler & Tanner Ltd, Frome and London

Front cover photo: A convoy of *Ptarmigan* vehicles in Southern Iraq at the end of the war. *(Soldier)*

To Anna and Joan

Foreword

Major General W J P Robins OBE
Assistant Chief of Defence Staff (Command, Control, Communications and Information Systems)

Mike Rice and Tony Sammes cover an aspect of the 1990/91 Gulf War all too frequently ignored: the underpinning of the operation by some determined and lively command and control system professionals. Many of the people in this book made prodigious efforts to meet the unprecedented demand for CIS for 1st Armoured Division in the desert and for the UK line of communications. They succeeded, just in time and with some fragility, from what was practically a standing start.

This book is not only a tribute to these unnamed professionals. There is much to learn from it. It should be read by all concerned with this vital area of military capability: the commanders and staff users, the service sponsors who ensure an integrated capability, the systems divisions who define operational requirements, the Procurement Executive who manage the projects and provide the interface with industry, the contractors who build the systems, and, finally, all students of this exacting and demanding field. We learnt a lot on Operation GRANBY. In the next crisis we are unlikely to be given the time to relearn it.

Acknowledgements

The authors particularly wish to extend their grateful thanks to the following people without whose contributions this book could not have been written: Tony Boyle, Colin Brown, Alan Carroll, Bob Cook, George Curtis, Duncan Galloway, Mike Galloway, Dick Hood, Stuart Kempster, John Lyde, Graham Paton, Fred Richards, Brydon Ritchie, Derek Robertson, Bill Robins, Dave Robson, Peter Sharpe, Paul Towers, Paul Vingoe, Peter Webster and Nigel Wood.

They also wish to acknowledge and thank the many people within the Ministry of Defence who have kindly provided advice, reviewed the manuscript and helped the authors to present a balanced view of the events that took place in the Gulf during Operation GRANBY.

The photographs reproduced as Figures 4.4, 4.5, 4.6, 5.3, 5.5, 5.6, 5.8, 5.13 and 5.15 are published by kind permission of Derek Robertson.

This work has been carried out with the support of the Procurement Executive, Ministry of Defence.

Contents

List of Figures

List of Figures

1.

Where We Came From

. . . the Gulf represents something of a sea change in the technology of war, perhaps as great as was experienced in the fourteenth century with the introduction of gunpowder;

Field Marshal Lord Bramall, 1991

From the start of the air war on 17 January 1991, television screens were filled with evidence of the use of the most sophisticated weapon systems ever employed. Briefings from the war zone were accompanied by video clips which made the attack on strategic targets deep inside Iraq with pin-point accuracy look no more difficult than playing a sophisticated video arcade game. One of the most telling newspaper pictures in the first days of the war was an artist's impression of the BBC's John Simpson looking out of the window of his Baghdad hotel room as a *Tomahawk* missile flew unerringly down the centreline of the road outside, *en route* for a target in the city centre.

The sight of *Patriot* missiles streaking across the sky over Riyadh and Dhahran to intercept incoming *Scuds* was an almost nightly spectacle for a time, and came to be taken for granted in a way which overlooked the incredible technological achievement it represented. Less visually spectacular, but certainly no less a technical achievement was the co-ordination of the air war, involving fighters, bombers, electronic-warfare aircraft for the suppression of air defence, reconnaissance aircraft and air-to-air refuelling tankers, controlled from the air by AWACS aircraft and from the ground.

We also saw much evidence of the very close co-ordination of air and sea power. Carrier-borne aircraft joined in the air battle alongside their ground-based counterparts, sharing the same refuelling tankers and control structure, as a totally integrated part of the attack plan. Action information systems on board ships of the allied navies co-ordinated attacks on Iraqi vessels with considerable success. All in all, it would appear that the tremendous array of awesome, high-technology firepower, with a few exceptions, performed very much 'as advertised'. What has yet to be seen is the extent to which technology – specifically information technology – has contributed to the campaign in terms of the overall command and control of the forces deployed, particularly of the ground forces.

This book concentrates specifically upon the way in which communications and information systems were used in support of the command and control of the forces involved. Further, it is most concerned with land forces and is written from a UK perspective (whilst recognising the high degree of integration of land, sea and air forces, and of the forces of some 35 nations).

According to one's point of view, the Gulf War either came at just the right time for the British Army, or at completely the wrong time. Given the degree of uncertainty and air of general change following the collapse of Communist regimes in eastern Europe, the fighting of what could have turned out to be a protracted war in the Middle East might have appeared to be a somewhat risky enterprise. On the other hand, had the situation in the Middle East arisen a few years later, the response to the ending of the Cold War in terms of force level reductions might have been seen as eroding Britain's capability to respond. A third viewpoint, which links the other two, is that the Gulf War was just what the Army needed to show that it still had a vital role, and persuade politicians not to extract too large a 'peace dividend' from the changes in Europe.

To appreciate the role played by CIS (command, control, communication and information systems) in the war, it is important to have an understanding of the CIS – doctrine and procedures, as well as systems and equipment – in place before the war started. An understanding of the influences which shaped today's command and control systems is, if anything, more important than simply knowing what their characteristics and capabilities are. There are obviously many influences, some conflicting, on the composition and equipment of a nation's armed forces, but the principal influence clearly stems from the purpose which the forces are required to serve in support of national foreign and defence policy. This will dictate both the geographical area in which the forces are to serve, and define the threat they are intended to counter.

Although the British Army had a considerable presence in the Middle East in the early post-Second World War period, it is long enough since that presence ended for its influence on the design of today's equipment to be largely ignored. More significant is the experience of individuals who have continued to serve in the Middle East, mainly on secondment and loan service appointments, such as those with the Saudi Arabian National Guard Communications (SANGCOM) project, and with the Omani Army. The principal roles of the Army are, and have been since the early 1970s:

- Defence of the United Kingdom – Military Home Defence (MHD)
- Contribution to NATO forces on the European mainland.
- Support of Government policy in maintaining security in Northern Ireland.
- Commitments out of the NATO area – Out of Area (OOA)
- Military Aid to the Civil Authorities (MACA)

Of these roles, those with the greatest influence on CIS doctrine and equipment have been, in order of importance, the NATO role, OOA and Northern Ireland.

The NATO/Cold War Legacy

The NATO strategy has been one of deterrence, that is, to deter an invasion of western Europe by the forces of the Warsaw Pact, led by the Soviet Union. When deterrence succeeds, as NATO's apparently has for over 40 years, wars are *not* fought. Hence whole generations of commanders have won their spurs through the direction of paper forces against a notional enemy, from their armoured command posts hidden under camouflage nets in wet beechwoods in the Teutoburgerwald, in rurally fragrant Westphalian farmyards, or inside derelict factories in the post-industrial Weser and Ruhr valleys. In fairness, of course, these paper wars have not been the sole source of experience. Many commanders have seen active service in a wide range of operational circumstances with Northern Ireland, the Falklands and Oman perhaps coming most readily to mind. We consider these and similar peace-time operations in later sections. Once a year a small part of the force has been allowed to exercise more realistically in the very different terrain of the Canadian prairie, whilst the major reinforcement and field training exercises such as CRUSADER in 1984 gave rare opportunities for commanders to practise the manoeuvre of actual forces on the ground rather than china-graph markings on the map.

The doctrine, tactics, organisation and equipment of the NATO forces were all based on the need to fight an essentially defensive, reactive war against a numerically superior enemy. For a long time, NATO's trump card was perceived to be the superior quality of its equipment and men. In the late 1980s, the latest Soviet main battle tank and their combat aircraft, such as the MiG-29 *Fulcrum*, which dazzled the crowds at the 1988 Farnborough Air Show with its incredible agility, were evidence that the Soviets were rapidly closing the equipment quality gap. Also, the environment of the potential battlefield was changing. Increased urbanisation was changing the fabled 'good tank country' of the North-West German Plain into a network of potential strongpoints, all within easy anti-tank missile range of each other and capable of being held by small numbers of lightly equipped, mobile defenders. Into the evergreen argument over which was the best tank, and what should the UK choose to replace its fleet of ageing *Chieftains* entered a new dimension – do we need a tank at all? Envious glances were being thrown at aircraft such as the US AH-64 *Apache* with its powerful anti-armour capability. The population of Germany in the late 1980s was becoming a lot less tolerant of the impact of NATO training on its environment, with ever stricter controls being placed on low flying and the field training of armoured formations.

The consequence for information systems of the NATO defensive doctrine was to place a huge emphasis on the need for commanders to be provided with accurate and timely information. When faced with outnumbering forces, who possess the offensive initiative, information is vital to enable forces to be concentrated in the right place at the right time to use them to best effect. Hence the emphasis falls on all the sub-systems of the basic command and control loop. Surveillance and target acquisition systems with ever increasing range, resolution and ability to work around the clock under any weather conditions were developed to service the performance monitoring sub-system.

In order to make sense of the vast amount of information flowing up the command hierarchy from the huge array of sensors, the emphasis in the decision-making sub-system was on data-fusion techniques, to correlate data from different sources into as coherent a picture of the battlefield as possible. Rapid, reliable and secure communications systems of adequate capacity were required to service both the performance monitoring and action sub-systems.

For an attacking force, able to choose both the time and place of its attack, the command and control system requirements are not so demanding, which may invalidate, at least in part, some of the comparisons and assessments of whether or not information systems worked as they had been intended to in the Gulf War.

One important legacy of the NATO experience for DESERT STORM was that of interoperability. NATO standardisation, though always fraught with timing and industrial problems, has produced results. British and American communication systems were able to interoperate, to significant tactical advantage. Also, the fact that British and American troops had trained together in NATO was very important in establishing a base of mutual understanding.

The 'standard NATO' threat briefings always used to start by saying that the Soviet attack would open with massive air strikes aimed at airfields, headquarters, command and control facilities, nuclear delivery means and ammunition and fuel dumps, accompanied by huge artillery bombardments. The ground attack would start with a number of feints designed to deceive us as to the direction of the main thrusts . . .

Perhaps we learned something else from all those years in Germany.

Out of Area (OOA) Operations

In operations out of the NATO area flexibility is the key word, since by definition it is not possible to predict in advance precisely where out of the area the next operation will take place. There is a whole series of contingency plans in place – the Joint Theatre Plans (JTPs) which deal mainly with the actions which would be required in order to evacuate United Kingdom citizens resident overseas who were threatened by situations in overseas countries arising through coups, insurrections and the like. Of the major OOA operations in recent years – the Falklands War (Operation CORPO-RATE) and the Gulf War, JTPs have been of little relevance, apart from providing templates for the immensely complicated plans required to schedule the movement of the required forces from a wide variety of starting positions to their destinations in the theatre of operations.

The influence of the experience of planning and executing OOA operations is felt in two key areas: rear link (i.e., overseas theatre of operations back to

UK) communications, and the establishment of Joint Force Headquarters (JFHQs) to command the operations. It was in the Falklands conflict that satellite communications (SATCOM) first proved their worth, as 'the only way home'. The advances made in this field since 1982 are impressive, to put it mildly.

Whereas in the Falklands journalists had to queue up to file their copy via the Army's one satellite link, during the Gulf War the networks had their own portable satellite stations. Many of the most potent images of the war came live on CNN from Baghdad, especially at the start of the air war, and later when CNN were the only remaining 'accredited' Western TV network in Iraq. It was a team from CBS, with their own satellite link, who entered Kuwait City many hours ahead of the first main echelons of the Coalition forces. In parallel with SATCOM developments there has been investment in improved high frequency (HF) long-range communications links, using techniques such as automated ionospheric sounding to find the best frequency to use from minute to minute.

OOA operations tend by definition to involve all three services, if only because of the distances entailed and the consequent need for secure air and sea transport. Joint, that is tri-service, planning and control has been applied to OOA operations for some time, under the direction of the United Kingdom Commanders-in-Chiefs' Committee (UKCICC), based alongside Headquarters United Kingdom Land Forces (HQ UKLF) at Wilton, outside Salisbury. The setting up and operation of Joint Force Headquarters (JFHQs) is practised regularly on training exercises. Doctrine and procedures have been devised, refined and documented in Joint Service Publications (JSPs).

Northern Ireland

Since the start of the present round of 'troubles' with Irish Republican and other terrorist organisations in Northern Ireland in 1969, the British Army has seldom had less than the equivalent of an infantry division deployed in the province, and at times, much more than that. There has been surprisingly little 'technology transfer' from the Northern Ireland operational experience to more conventional areas. Amongst the reasons for this one must count the fact that Northern Ireland has been seen as something different from the mainstream role of the Army, and a temporary situation.

Much use has been made of ingenious technology in an effort to gain the edge over the terrorists. That technology included the early use of computers in support of both operational and intelligence functions. Few would argue that the key to combating terrorism is high quality intelligence, and the application of computers to the management of the large amounts of information required was realised early in the campaign.

On the communications front the emphasis was on low-scale, personal communications as befitted the nature of the anti-terrorist operation. Advances in similar communications systems, such as those developed for the civil emergency services, have been exploited by the Army to good effect. At the human level, the main impact of Northern Ireland operations has been in the training of a generation of junior commanders. The key unit of deployment on the streets of Ulster is the half-section 'brick', commanded by a junior non-commissioned officer (JNCO). The degree of responsibility carried by the JNCO under operational conditions in Northern Ireland has greatly influenced the development of their self-reliance and all-round leadership qualities, and led to Northern Ireland's being dubbed the 'Corporals' War'.

The Influence of Technology

There are some interesting paradoxes at work in the way technology has been applied to battlefield CIS. For many years, military research and development was very much at the 'leading edge'. It would be fair to say that in CIS, many concepts were first considered long before the technological capability existed to implement them. The *Ptarmigan* area communication system is a good example of this. The basic concepts of an area communication system, whose deployment is independent of the chain of command, were first elaborated in the late 1950s. At that time, the hardware technology and software techniques which would have been required to put a system of adequate performance and reliability into the field simply did not exist. An important point to note at this stage is the 'technology transfer' between military and civil areas.

Considerable benefit from the multi-million pound investment in military CIS technology migrated to the civil communications area. This is one of the prime causes of the paradoxical situation which now exists. Because of the much faster product development cycles in the civilian field, by the time military systems which had been state-of-the-art when originally conceived, come into use, they have all too often been overtaken in capability by their civilian counterparts. *Ptarmigan* stands up quite well to such examination, but in truth provides little more than the user of any modern, civilian telecommunications service has come to expect. Computer-based information systems stand up much less well.

The user of office systems, based on networked personal computers with sophisticated graphical user interfaces and a comprehensive array of well-designed, integrated software packages, is somewhat dismayed by the pedestrian performance of the latest operational information systems. There is therefore an expectation gap on the part of the users; only the most enlightened of them are able to understand the different circumstances under which tactical systems are procured, and the difficulties of the environment in which they must operate, and are prepared to put up with their shortcomings.

The Demand for More Information – the New Management Strategy

On 1 April 1991 a profound change occurred in the way the Ministry of Defence and the Armed Forces are managed. New Management Strategy

(NMS) is the MOD's answer to the Financial Management Initiative (FMI) brought into central Government in the mid 1980s. The basic principle of NMS is to make commanders accountable for the money spent in the discharge of their responsibilities. NMS demands that commanders know how their resources are being used and the extent to which objectives are being met. This has created a new demand for information, to be satisfied by a whole range of new systems.

Information Systems (IS) Strategy

At about the same time that Saddam Hussein's forces invaded Kuwait in August 1990, the first comprehensive Army Information Systems Strategy Study presented its final report. The review had been carried out against a background of all the factors outlined above, and in its later stages with the threat of impending war in the Gulf to be taken into consideration. To limit the area of study to a practicable size, the initial Army IS Strategy had excluded battlefield systems. There had also been inevitable constraints of affordability which had ruled out some of the more ambitious and far-reaching options so that, in some cases, capability enhancements were contingent on there being corresponding efficiency savings.

An important result of the Army IS Strategy was the creation of a target information systems and technology architecture, which envisaged the introduction of an IT (information technology) infrastructure at unit, headquarters and MOD level through which the necessary IS applications could be provided.

Systems in Place

In mid-1990, very little of the IS infrastructure was in place. At unit level, apart from a microcomputer-based personnel administration system introduced in the early 1980s and a number of 'private' microcomputer systems, there was very little support. Microcomputers were starting to appear in brigade and divisional headquarters. One division in particular (3rd Armoured Division) had gained some useful experience through pioneering the use of portable computers which were used to support staff work and office procedures both in barracks and when deployed. At Headquarters 1st British Corps in Germany a headquarters-wide information system (CORMIS) had been installed, and was growing as locally developed applications were added to the system, but its evolution was hindered by a shortage of resources.

Further west, at Headquarters British Army of the Rhine (BAOR), a comprehensive office automation/information system had been in use for a number of years. This ICL-supplied system was called CALAIS (Command and Logistic Information System). Over in the UK, at Headquarters UKLF, there was a system called MAPPER which was based on proprietary hardware and software. The wide area communications infrastructure to support the fixed, 'non-operational' CIS was a combination of rented circuits and both MOD-owned and rented equipment in the UK, with a substantial MOD-owned bearer network in Germany.

In the operational, battlefield area the major strength was the success of the *Ptarmigan* trunk communications system, which was reaching a useful degree of maturity some four to five years after its initial deployment. The system provides secure telephony, telegraph, facsimile and data transmission for users at headquarters from corps to brigade, and a cellular radio-like mobile service via the Single Channel Radio Access (SCRA) sub-system, deployed down to unit level. Less suitable for deployment was the *Wavell* Command and Control Information System (CCIS), deployed at headquarters from corps to brigade, using *Ptarmigan* as its communications bearer. For the Royal Artillery, their long wait for BATES – the Battlefield Artillery Target Engagement System – was coming to an end, with the first issues of the system at regimental and battery level. Systems required to support other Battlefield Functional Areas – Air Defence, Intelligence and Electronic Warfare and Combat Support/Logistics, were still some way from the field, although the logistics area was served by a number of systems of varying degrees of maturity and utility.

Approach to the Book

A number of different approaches to this book were considered. They ranged from a chronological, purely technical account of what happened, to one based much more on the human aspects, and *why* things happened the way they did. Both extremes have their merits, and the current structure is a compromise between them. Our objective in writing the book was to demonstrate how the CIS principles and techniques described in our text-book,[1] were applied in a real situation, to draw some conclusions and present some lessons for the future. Hence the facts of what was done, and some comparison between those facts and a theoretical, ideal, textbook situation must form the book's essential core. At the same time, we have to recognise that CIS embraces people and procedures as well as technology and equipment. The facts about what CIS was available to the forces in the Gulf, how it was used and the impact it had on the outcome were shaped as much – if not more – by the influence of the people, organisations and procedures as by technology. We could not tell the whole story, let alone draw any conclusions from it or present any lessons for the future without giving a proper account of why and how the situation developed in the way it did.

We have agonised at length over the style and over the right way to introduce the many people who contributed to the story. Ideally, we would have liked to mention them all by name, quite openly, but had some difficulties with that approach, not least that all who were interviewed gave their very

---

1 *Communications and Information Systems for Battlefield Command and Control*, M. A. Rice & A. J. Sammes, Brassey's, 1989.

frank views quite freely without needing to be concerned about either security classification or political sensitivity. Resolving the latter problem for an unclassified, public domain publication was left to the authors and, of course, the good offices of the Ministry of Defence. As many of those interviewed are still serving, we considered that we had a duty to preserve their anonymity, even it that denied us the chance of giving credit where an enormous amount of it is undoubtedly due. In order to satisfy the latter condition yet maintain a degree of 'reader friendliness' we attempted at one stage to give all the 'players' pseudonyms. When we read the results of that experiment, it just did not seem to work, so we resorted to the use of appointments in describing the people. If that seems a bit cumbersome at times, we apologise and hope it will not distract unduly from what we believe is an important story for anyone with an interest in the application of modern technology to warfare, and essential reading for anyone with a professional involvement in that field.

2.

Where We Should Have Been?

The toe of the star-gazer is often stubbed.
Russian proverb

Introduction

A number of books have been published which describe the events of the Gulf War from a general, politico-military viewpoint. It is not our job in this book to cover again that ground in any great detail, but to set our specialist account of the conflict into perspective, we must at least present a resumé of the war itself and the events leading up to it.

In setting the scene and describing the preceding events, we have taken the opportunity to go further and outline a hypothetical, theoretical response to those events. We have done this for two reasons; first, to extend the arguments presented in our textbook and show how we believe a force such as that selected to participate in the Gulf War, and the whole process of assembling, mounting and sustaining that force, should have been supported by communications and information systems (CIS). The second reason is that the 'ideal' picture painted in this chapter provides the reader with a yardstick against which to measure the quality and extent of the support which was actually provided, as its description is unfolded in the subsequent chapters. The difference between the ideal and that actually achieved is a reflection of a number of factors including the fact that the UK's contingency plans had not envisaged such a large out-of-area deployment and the constraints of the time required to develop, procure and bring new systems into effective service.

The Beginning of the Conflict

On 2 August 1990, Saddam Hussein's troops invaded Kuwait, in the culmination of a period of threats and increasing tension which had been going on for most of the preceding month. The pre-dawn thrust into Kuwait City by Iraqi tanks put an end to the 'will they, won't they' speculation of late July and forced consideration of a response by the West, under the banner of the United Nations. The urgency of a response was driven by the fact that Saddam Hussein undoubtedly had the capability to move south from Kuwait and on into Saudi Arabia. The signs and wider political circumstances before 2 August had not been considered sufficient to justify international action to pre-empt the invasion of Kuwait and those countries making up the Coalition were now determined to send Saddam Hussein a clear message that any further moves would not be tolerated.

As details of the package of economic sanctions and an embargo of Iraq were still being worked out, a massive airlift of US forces into Saudi Arabia began. Britain reinforced its standing naval patrol in the Gulf area with the despatch of *Tornado* air defence aircraft to Saudi Arabia and *Jaguar* ground attack aircraft to Oman. Four weeks later, on 14 September, the British Government announced the commitment of ground forces to the Coalition, in the form of 7th Armoured Brigade plus supporting troops. With the passing of UN Resolution 678 on 29 November 1990, the Coalition was authorised to use all necessary means to recover Kuwait if Saddam Hussein had not, by 15 January 1991, satisfied the previous UN resolutions setting the terms for a peaceful solution to the crisis.

Shaping the Force Structure

There is a technique taught in the British Army, and probably in most other armies, for the assessment of a situation and the determination of the most appropriate course of action to deal with it. This is the Appreciation, and it is as relevant to the determination of how best to carry out a section attack as it would be to the resolution of how to restructure the Army following the political and military reshaping of eastern Europe and the Soviet Union. The latter exercise would occupy many staff officers for a long time and produce an output covering many acres of paper, whilst the former would be a corporal's mental process, backed up by a few scribblings in a field notebook. The essential process is the same; what is the aim we are trying to achieve? What are the factors affecting our achievement of it? What options are open to us, and finally, which one is most likely to bring success? In the tactical appreciation, factors are considered under four headings:

- Ground
- Enemy
- Own forces
- Time and space

In the more complex, politico-military appreciation there are, of course, other factors to consider. It will suit our purpose in tracing the origins of the UK contribution to the Gulf War to concentrate on the simple factors of the tactical appreciation.

Ground

Ground has to be considered at two levels; first, from a global point of view, that is, taking a wide enough view to encompass not only the theatre of operations in the Gulf, but also the areas from whence the forces and their supporting *matériel* would come. Secondly comes the more conventional assessment from a tactical point of view of the ground in the projected area of operations.

Geography

The distances involved both between the UK and the Gulf, and also within the Gulf theatre itself are considerable. Typical flight time for transport aircraft from airports in the south of England to north-eastern Saudi Arabia is around six hours. The journey by sea, which had to be taken by most of the vehicles and equipment once the decision had been taken to deploy an armoured formation, takes 21 days.

The link between the wider, global view and the more local, tactical view is the location of the ports of entry for men and *matériel* arriving by both sea and air, for their location defines the starting points for the internal lines of communication. Clearly, it is important to establish a mounting base which is as near to the intended area of operations as security, and the availability of suitable facilities, will allow. The port of Al Jubayl, on the Gulf coast of Saudi Arabia some 150 miles south of Kuwait City, was chosen to be the Force Maintenance Area (FMA). During GRANBY 1 (the deployment of 7th Armoured Brigade and supporting troops) alone, some 9,000 passengers, 5,000 vehicles and 1,200 containers holding 23,000 tonnes of ammunition and general cargo entered the theatre of operations through Al Jubayl.

Terrain

The whole of the area of concern is described as a desert plain. The terrain is rocky, with varying depths of sand, and no significant relief features. Some of the most significant obstacles to movement are oil pipelines, such as the Trans-Arabian Pipeline (the Tapline). The pipe is about 80 cm in diameter and is a serious obstacle to both wheeled and tracked vehicles, except at the specially constructed crossing points. The relative abundance of oil pipelines and other installations in the eastern littoral approach to Kuwait City from the south was one of the reasons which prompted the British commander, General Sir Peter de la Billière, to secure the re-assignment of 1st (British) Armoured Division to take part in the 'wide left hook' with 7 (US) Corps, out in the hard, flat desert to the West of Kuwait.

Climate

Although the Gulf region is, on average, one of the hottest areas on earth, there are significant differences between mean summer and winter temperatures. During the summer the mean daily temperature over the whole of the region is over 30°C and maximum day temperatures of around 50°C have

FIG 2.1 The Gulf Area of Operations

been recorded. By contrast, the average daytime temperature in Kuwait in January is only 16°C, and night-time temperatures approaching freezing point are not unknown. Further inland, at Riyadh in Saudi Arabia, the night-time temperature can fall to well below 0°C. From a strictly military view-point, it is the extreme maximum temperatures which pose the most serious threat to men and equipment. The effectiveness of soldiers when clad in full NBC protective equipment falls off rapidly as the temperature goes up. In the typical summer daytime temperatures of the Gulf, just existing, let alone fighting in full NBC kit, is a painful and debilitating ordeal. The maximum temperatures also come dangerously close to the tolerances specified for military electronic equipment. When the equipment, which is itself generating heat, is enclosed within armoured vehicles, sealed for NBC protection, the likelihood of exceeding the high temperature specification is high.

The Gulf region is popularly seen as a place where it seldom, if ever, rains. It is, in fact, officially classified as 'arid', having a mean annual precipitation of less than 200 mm. As members of the Operation GRANBY task force can testify, most of that rainfall appears to fall at the same time. During the winter months, heavy downpours are fairly common. They are often extremely localised, and very unpredictable.

The climate had a most important effect on the timing of the war. The severe difficulty of mounting operations during the summer months imposed an effective deadline on the start and duration of the ground war. This was allied to the problem of fighting, with Muslim allies, during Ramadan, the Islamic period of religious observation when Muslims fast during daylight hours.

CIS-specific factors

The physical environment had a number of significant influences on the provision of CIS support. Obviously, the geography of the situation dictated the area and distances to be covered by communications. The distances in-theatre were considerable, particularly when compared with those over which British forces would have expected to operate in NW Europe. A Forward FMA was established some 350 km north-west of the FMA at Al Jubayl, along the Tapline Road (named as Main Supply Route [MSR] DODGE). Once the ground offensive began, the Division covered 480 km in 100 hours, adding the complications of sheer speed of movement into enemy-held territory to the ordinary difficulty of mere distance.

The communications medium most affected by the terrain and the distances involved was the multi-channel radio relay used to connect nodes of the *Ptarmigan* trunk communications system. In the Saudi Arabian desert terrain, radio relay path lengths of several tens of kilometres were the norm. Some six links were needed to establish a radio relay chain from Al Jubayl to the FFMA. Quite apart from the technical fragility of such an arrangement, it consumed an inordinate number of men and amount of equipment, and posed problems of security for isolated detachments. As we shall see in a later chapter, the problem was overcome through the use of satellite communications.

The Enemy

The total strength of the Iraqi Army before the Gulf War was estimated at around one million men. During their eight-year long war of attrition with Iran, the Iraqis are believed to have lost upwards of 105,000 men killed. It is hardly surprising that the quality of those men under arms in mid-1990 was variable, to say the least. The highest calibre troops were to be found in Saddam Hussein's Republican Guard, whose loyalty was encouraged through special privileges. The quality of the other troops was generally on the low side, with extremes of age not usually encountered in armies elsewhere.

Much of the equipment used by the Iraqi forces originated in the Soviet bloc, but some came from the West, including small quantities from the UK, which at least made sure that we had a good idea of its potential capabilities in trained hands.

Of specific note from a CIS viewpoint was the Iraqi capability in command and control, and electronic warfare (EW) (or counter-command and control). In cultural and procedural terms, the Iraqi armed forces relied heavily on centralised command, with Saddam Hussein himself as the commander-in-chief. Little initiative was allowed to subordinate commanders. Communication from the top down through the levels of hierarchy was therefore important, and one of their strengths was the way in which use was made of alternative means of communication to provide a degree of resilience. This included everything from microwave radio relay through High Frequency (HF) radio to landline and physical courier.

The Iraqi forces were known to have a Very High Frequency (VHF) intercept, direction finding (DF) and jamming capability. In the face of this threat it was important for the coalition forces to have effective defensive EW procedures, techniques and equipment.

Own Forces

Faced with the prospect of assembling an armoured formation to send to the Gulf, the MOD's planners had little choice but to call on the resources deployed in the British Army of the Rhine (BAOR), where Britain had some 55,000 troops stationed, organised into a corps (First British Corps, or 1 (BR) Corps) of three armoured divisions and supporting troops. The force eventually deployed to the Gulf was nominally one of the three divisions (1st Armoured Division). This took most of the equipment of 1 (BR) Corps and many additional men to bring it up to a state of readiness capable of the enhanced mobility required of a division operating over long distances in an offensive role. The larger part of the Corps' artillery, communications and logistic support was also committed to 1st (British) Armoured Division in the Gulf.

It was known from the outset that the UK contribution to the Coalition forces would have to interoperate with forces of other nations, particularly those of the United States. This was advantageous because of existing NATO standardisation agreements, experience of co-operation on Exercise REFORGER (REturn of FORces to GERmany – from the continental United States) and other major training exercises. On the communications side, there was a good deal of experience of UK/US co-operation. The interoperability between *Ptarmigan* and its US counterpart, (MSE – the Mobile Subscriber Equipment) had been demonstrated in Germany. It should be stressed that this experience of co-operation was between the British Army and elements of the US Army committed to NATO; the US Marine Corps, under whose command the UK contingent was placed initially did not have the same experience. This was one of the reasons for the redeployment to come under command of the US VII Corps, a unit drawn from US forces in Europe.

Time and Space

The time considerations appeared to fall into two stages; first, the need to assemble as soon as possible a force which would be sufficient to deter any further territorial incursions by the Iraqi forces, that is, into Saudi Arabia. Once this had been achieved, and it became clear that peaceful means were not working, the emphasis shifted, backed by Resolution 678 in the UN, to the threat of offensive action to eject the Iraqis from Kuwait.

The most immediate problem was that of how to assemble the force, which, as we have said, was drawn mainly from Germany, and move it to the Gulf. Not just the actual men and vehicles comprising the two armoured brigades and their supporting troops, but the thousands of tons of supplies needed to sustain them. A very important role was played from the outset and throughout the campaign by the logistic staff, especially those organising the transport and movement of personnel and equipment both to and within the Gulf theatre.

The first stage started within a few days of the invasion, with the deployment of air forces to the Gulf. The decision to deploy ground forces, in the shape of 7th Armoured Brigade was announced on 14 September 1990, and the first ships to take the force to the Gulf were scheduled to sail just two weeks later. A problem facing the planners was the complete absence of contingency plans to deploy a heavy armoured formation Out of Area, and the time taken to stabilise the Order of Battle – ORBAT – of the force, that is, its detailed composition in terms of men, vehicles, weapons, equipment and supplies.

Time played a major part in planning for the offensive option. The essence of the problem was that an adequate force had to be assembled, acclimatised and worked-up in theatre in time to allow action before two major deadlines; the onset of hot weather and the Ramadan period. There were powerful arguments for starting and completing any offensive before these deadlines. As indicated earlier, this placed a premium on completing the operation before mid-March.

The Force Structure

The initial decision was to deploy 7th Armoured Brigade reinforced by much heavier than normal support. The ORBAT of the force deployed for Operation GRANBY was:

- Headquarters 7th Armoured Brigade and 207 Signal Squadron
- An armoured reconnaissance squadron, equipped with the *Scorpion* Combat Vehicle Reconnaissance (Tracked) – CVR(T), – 'A' Squadron, The Queen's Dragoon Guards
- Two armoured regiments, equipped with *Challenger* main battle tanks, The Scots Dragoon Guards and The Queen's Royal Irish Hussars
- A mechanised infantry battalion, equipped with the *Warrior* Mechanised Infantry Combat Vehicle, 1st Battalion the Staffordshire Regiment

- An artillery field regiment equipped with 155-millimetre self-propelled howitzers, 40 Field Regiment
- An air defence battery
- An armoured engineer regiment, 21 Armoured Engineer Regiment
- The logistic support forces, organised under a Brigade Maintenance Area (BMA) headquarters, were:

  A divisional transport regiment
  An armoured field ambulance
  An ordnance battalion
  Elements of Royal Electrical and Mechanical Engineers armoured workshops
  A provost detachment
  A Postal and Courier detachment

- A Force Maintenance Area was established, comprising:

  An FMA Headquarters with communications elements
  A *Ptarmigan* squadron from 1 Armoured Division HQ and Signal Regiment
  A Royal Engineer construction company
  Elements of 14 Topographic Squadron Royal Engineers for survey tasks and map supply
  A Postal and Courier detachment
  Elements of a general transport regiment (10 Regiment Royal Corps of Transport)
  A tank transporter troop
  A field hospital
  An ordnance battalion
  Remaining elements of the armoured workshops
  A provost company.

The GRANBY force totalled around 9,500 troops, of which some 4,500 were in the combat 'teeth', about 2,000 in the immediate logistic support to the brigade (ie under the BMA), and the remaining three thousand in the FMA 'tail'. This figure is about twice the size of a 'normal', i.e., BAOR-based armoured brigade. In addition, there were some 2,000 medics who were part of the same deployment. In some areas, the support provided for the brigade equalled to that more usually provided for a whole armoured division, and in others it was nearer to that available to all of 1(BR) Corps in the Germany scenario.

The size of the force committed to Operation GRANBY was increased on 15 November to an Armoured Division (1st Armoured Division) with two Armoured Brigades, 4th and 7th.

In considering our ideal CIS arrangements below, we have taken the decision to deploy a heavily supported armoured division as our basis. In reality, it was not a clear-cut decision based on the factors available from the outset, but the result of a process of escalation. Many of the decisions taken and arrangements made when the force was to be based on a brigade were no longer valid when the force's size was increased, particularly as regards the CIS.

CIS Support within the Force

It is at this stage that we start to enter the realms of the theoretical, ideal 'textbook' situation. We have traced the arguments and factors which led up to the commitment of a full armoured division. We shall proceed with our dissertation on the ideal way to support such a force on the basis that the decision to deploy 1st Armoured Division was a homogeneous process based on the factors of the situation, and ignore the complication of the earlier planning assumption set with the deployment of 7th Armoured Brigade.

So, given that the task is to deploy a heavily supported armoured division to operate in an arid climate, in desert terrain, over long lines of communication, in co-operation with forces of both NATO and non-NATO allies, many thousands of miles from home against a numerically superior, well-equipped enemy, how should its CIS support be organised and provided?

The principles we have followed in answering the question above are, in the main, those presented in our textbook. Everything we describe below could be provided with today's (and in many cases, yesterday's) technology, embodied in tried-and-tested equipments and systems. Our description is based on the five battlefield functional areas (BFAs) of manoeuvre control, intelligence and electronic warfare, air defence, fire support and logistics, and the communications infrastructure needed to support them. Interoperability between the BFAs, and between the UK force and other Coalition forces, is also addressed.

Manoeuvre Control

Manoeuvre control is essentially the G3 Operations function. CIS support to this function includes the presentation to the commander of a clear, up-to-date picture of the overall tactical situation and the means to formulate and disseminate orders. The former requires information on both own forces and those of the enemy, as well as the 'neutral' factors of ground and weather to be presented as a single, coherent view of the situation on which the commander can base decisions for future action.

The degree of integration between systems and fusion of information required by the total situation display is just about possible with the current state of the art, but has not yet been implemented in any practical form on the battlefield. In most systems, the fusion and integration functions are performed manually and the information provided to the commander in staff briefings is supported by displays on paper maps.

Although not strictly part of the manoeuvre control function, CIS support to the staff function in headquarters or 'tactical office automation' (OA) can achieve great improvements in the efficiency of the staff process, particularly the preparation of complex operation orders. Given that the ability to assess a situation and react to it 'within the enemy's command and control loop' is a key to success, then the contribution which can be made by CIS/OA in reducing reaction times is significant. We would therefore expect our force to be equipped with OA.

Position Finding and Navigation

Position Finding and Navigation is not strictly a manoeuvre control application, but plays a vital role in it. As it is a particular problem in the desert, it is well worth a mention here. Practically all information held by a tactical command and control information system is spatially related. Hence the information system needs up-to-date and accurate information on locations. That is essentially an information reporting and distribution problem – assuming that the units and sub-units have a reliable and accurate means of determining both their own locations and those of any activities they may report. A system introduced in the US Army in the late 1980s was designed to combine both the position locating and the reporting function. The Position Locating and Reporting System (PLRS) is a UHF radio system in which individual stations can compute their own locations based on the time of arrival of signals from other stations, and transmit their locations to a control station. The system's position-finding capability relies on the fixing of certain base stations by accurate survey. In the enhanced version (EPLRS), the radio system is also used for low to medium capacity data transmission.

The British Army has traditionally relied upon conventional navigational techniques based on map, compass, skill and experience. Inertial navigation systems have been used by artillery units and others requiring particular precision, but, like PLRS, these need calibration against accurate reference points. Conventional navigation and position finding in the desert are complicated by the relatively featureless nature of the terrain, and the poor reliability of mapping, especially in areas where the surface is constantly shifting.

The literally space-age solution to this problem comes in the form of the satellite-based Global Positioning System (GPS). Using calculations based on the time difference of the arrival of signals from a 'constellation' of low-orbit satellites, GPS can provide its users with a read-out of their location to the nearest 15 metres. Even allowing for the occasional brief pause in operation when fewer than the required number of satellites were above the horizon, GPS played a crucial part in enabling Coalition forces to advance rapidly across the featureless desert to the west of Kuwait.

Intelligence and EW

Before the deployed force was in contact with the enemy the principal sources of intelligence were outside the Division's control. Sources such as reconnaissance aircraft and satellites, electronic and signals intelligence

(ELINT and SIGINT) all had a role to play in piecing together the picture of Saddam Hussein's deployment and movements. As well as the general intelligence picture, two specific areas were of high priority for intelligence gathering. The first of these was the effect that the air campaign was having on the Iraqi forces and infrastructure. This information was essential in the process of deciding when the time was right to launch the ground offensive. The second area of special interest was a defensive concern – the location of *Scud* missiles and launchers. These Soviet-designed, free-flight rockets represented the only serious form of physical retaliation taken by the Iraqis against the Coalition forces. Their use against random targets in Israel was an attempt to divide the Coalition and to bring Israel directly into the conflict. For both reasons it was vital to the Coalition that the threat be countered in every possible way, from the detection and destruction of launchers and missiles on the ground to their interception and destruction in flight. This latter was accomplished with seemingly spectacular success through the deployment of the *Patriot* missile system to both the Coalition forces and to Israel. Subsequently, however, this apparent success of *Patriot* has been put in doubt by analyses undertaken in the USA which purport to show that, while the missiles had a good record of intercepting *Scud* rockets, they were not particularly successful in the crucial objective of destroying the warhead.

After the land battle had begun these strategic level sources remained vital but were then supplemented by assets organic to the Division, such as the armoured reconnaissance squadron, artillery drones and the information gathered from the increasing flow of Iraqi prisoners of war.

From a CIS point of view, what was important to the Division was the means to receive and collate all the information from this wide variety of sources and turn it into a useful and meaningful picture which could be presented to the commander. The remarks made earlier concerning Manoeuvre Control systems apply equally here. Although the technology certainly exists to automate the collation and presentation of information in a map-based, visual way, its widespread adoption and integration into procedures at the tactical level of command is yet some way off. It would be unrealistic to expect our force to have progressed beyond the manual collation and evaluation of intelligence information, with presentation via paper map overlays. Some means of transmitting graphical information between locations should be provided.

Fire Support

CIS support to the fire support BFA fills two main functions: at the local, fire-unit level, the necessary computation of target details to bring fire to bear accurately on the target, and above the local level, the co-ordination of fire from as many fire units as are needed to achieve a given mission, are available and within range.

The local, ballistic computation function was one of the first applications of computers on the battlefield in the form of the Field Artillery Computing Equipment (FACE). The techniques are well-proven and capable of satisfactory implementation on hardware of relatively modest power. Indeed, quite adequate results can be obtained by using nothing more sophisticated than a programmable pocket calculator.

The wider co-ordination function is more difficult to provide. It demands a careful analysis of the procedures involved in the whole fire-support system. Its implementation requires a widely distributed system, with elements covering the observation, fire direction and liaison functions, as well as the fire units themselves, all interconnected by reliable and secure data communications.

It would be reasonable to expect our force to be provided with a computer system which incorporated both local and co-ordination functions. The local function would perform its calculations on target data entered into the system at the point of observation and pass the results direct to the fire units. The system would contain algorithms for calculating the best combination of fire units to achieve a given effect on the target according to the commander's priorities and the fire units within range. It would be supported by a data transmission system, probably based on a combination of conventional combat net radio, packet radio and *Ptarmigan* SCRA.

It is worth noting that the Battlefield Target Engagement System – BATES – was not deployed. Although units which had been issued with BATES in Germany were deployed to the Gulf, the system had not yet been accepted by the MOD, and was left behind. The veteran FACE continued to be the mainstay of artillery computing, providing local ballistic calculations as described above. FACE, designed in the 1960s, was one of the first applications of computers on the battlefield.

Fire support also includes tactical air support. Although we have not described the air forces' contribution in any detail, all will be aware that air power was a major factor in the Coalition's eventual victory. Throughout the 'air war' which opened in spectacular style in the early hours of 16 January 1991, the offensive was carried out almost entirely by land- and sea-based aircraft and missiles. During this time the land forces continued their build-up, and acclimatisation and training programmes. Once the land battle was joined, air power continued to play a decisive role, but had to be carefully integrated into the overall fireplan, which now included both tube and rocket artillery, such as the Multiple Launch Rocket System (MLRS). Such integration demands a minute-to-minute knowledge of the availability of air resources and rapid, reliable communications at all levels down to and including contact between the pilots and the troops on the ground. International interoperability plays an important part here too, since it had to be possible within the Coalition for any aircraft to support any ground troops, for instance, US aircraft in support of UK ground troops, UK aircraft in support of Egyptian ground troops, and so on. We would expect our force to meet the air support CIS requirement by providing liaison officers with access to the air force mission planning system at divisional and brigade headquarters, and all the necessary communications support, including ground-to-air links.

Air Defence

The provision of effective defence against enemy air action calls for a combination of the means to detect and identify enemy aircraft, the means to destroy them, preferably before they have launched their attack against ground targets, and the means to command and control the whole process. This in turn requires rapid and reliable communications between the detection and the destruction elements of the system, and between both elements and the air-defence co-ordination centres at formation headquarters. Since enemy aircraft could theoretically appear from any direction, the air-defence system must be able to interoperate with flanking formations of whatever nationality.

Our force should be equipped with an air-defence system which is flexible enough to be able to operate with a wide variety of allied sensor and fire unit types and to make effective use of whatever communications medium may be available to it. It should provide the means to identify threats and track them across our area of responsibility, either deleting them as they are destroyed or passing them on to neighbouring air-defence systems.

A special form of air defence is that against the surface-to-surface missile threat. Though Iraq's use of the Soviet-designed *Scud* free-flight rocket was of little military concern to the Coalition, its use was a cause of significant wider concerns, not least in relation to Iraq's attempted use of the weapon to provoke an Israeli involvement in the crisis. The threat was heightened by our concern that the enemy might have the capability to equip their *Scud* missiles with chemical warheads. This is one example where, as will be told in a later chapter, the actual technology used far surpassed any 'ideal' situation we may postulate here. At this stage let it suffice to say that we would expect our force to be equipped with the means of detecting hostile missiles as early as possible in order to issue a warning to those in the potential target area, and as effective as possible a means of destroying incoming missiles.

Communications

The primary means of communication below brigade is combat net radio. For the immediate, second-to-second control of the contact battle there can never be a substitute for voice communications. For less immediate communications requirements, such as the submission up the chain of command of routine reports, situation reports, intelligence reports, casualty reports, etc., data transmission using pre-formatted messages has an increasingly important role to play. The radio system must be flexible – it must be easily reconfigured to take account of regrouping and changes in deployment and, above all, it must be secure.

Single Channel Radio Access (SCRA) provides an alternative link from battlegroups and some specialist sub-units back to brigade headquarters, as well as a connection into the force-wide trunk communication system.

Interoperability

Internal
The CIS used to support the different BFAs within our force would have the capability to interoperate and share information via a common user database. This is very much in line with work done during the 1980s in the UK to define a 'Goal Architecture' for battlefield CIS.

External
We would expect our force's CIS to be able to interoperate with those of our allies. Within BFAs, for example, we would expect our fire support and air-defence systems to be able to exchange information with those of our allies. It should be possible for a fire-direction centre in a neighbouring, allied formation to set up a fire mission which brings to bear guns on both sides of the interformation boundary. Likewise, it should be possible for air-defence fire units in our area to engage targets detected by a neighbour's sensors.

CIS Support to the Assembly, Mounting and Maintenance of the Force

The Problem

At the heart of the problem is the time and space situation outlined previously. It would have been a complicated enough problem simply to move the force from a UK base and to sustain it from there. It was made much more complicated by the fact that much of the force and a large amount of its equipment and stores were already deployed in Germany in their Priority 1 NATO role. Hence men and equipment had to be moved from Germany to the Gulf, under control of Joint Headquarters in the UK.

It was of some consolation that tried and tested communications and information system links exist between Germany and the UK. The Headquarters British Army of the Rhine (HQ BAOR) Command and Logistic Information System (CALAIS) has cross-Channel links. Also, the Royal Army Ordnance Corps stores system, centred on Bicester in Oxfordshire, is linked to forward depots in Germany. Hence there was already in place a considerable CIS infrastructure between BAOR and the UK, although it was far from ideal in many respects. For example, there was no compatibility between CALAIS and the Headquarters UK Land Forces MAPPER system, nor with the RAF Air Staff Management Aid (ASMA) which was adopted as the Force level CIS. Additionally, there were difficulties in communicating between HQ BAOR and the CIS at Headquarters First British Corps CORMIS system.

Given that Operation GRANBY was an OOA operation, we would expect to see the contingency OOA communications links deployed. These are a satellite communications (SATCOM) link with high frequency (HF) radio back up to the force mounting base in the overseas theatre and to the Joint Force Headquarters (JFHQ) once established. As we shall see, the scale of the operation demanded a much greater strategic level communications capacity than this would have provided.

3.

Where We Got To

During the first 90 days of the military buildup, we put in more communications connectivity [in Saudi Arabia] than we've had in Europe during the past 40 years.

USAF Lt. Gen. James S Cassity, 1991

Introduction

We have to this point examined the CIS infrastructure that we might reasonably expect to have seen used by the British Army in the Gulf given our understanding of CIS fundamentals and our knowledge of the existing Priority 1 CIS systems. In this chapter we give a brief overview of the CIS systems that were actually put in place for Operation GRANBY. Why they did not turn out to be as predicted and what really happened is the story of the following three chapters.

Communication Systems

Tactical Communications

The three main tactical communications systems that were used in the Gulf are not at all unexpected.

Ptarmigan provided the trunk communications network system with Single Channel Radio Access (SCRA) centrals deployed well forward. *Clansman* Very High Frequency (VHF) Combat Net Radio (CNR) was used as the primary means of command from Division down to combat teams with *Clansman* High Frequency (HF) radios providing a backup communications facility. In principle, there was nothing unusual about any of this.

The important differences, however, between this operation and BAOR working were those of scale and the limited advanced civilian communications infrastructure in the theatre of operations. Although only a reduced size division of two brigade groups was deployed, the distances involved in the Gulf were so great that it required almost all of 1 (BR) Corps' communications assets to meet the requirement. Even so, the *Ptarmigan* trunk nodes were so over-stretched that it was necessary to devise 'satellite bridges' to link parts of the *Ptarmigan* network together.

Combat Net Radio suffered because of a lack of sufficient rebroadcast stations and their inability to keep up during the very rapid advance, likened by some to the Paris-Dakar rally. There was not the option, as had always been the case in BAOR, to supplement both tactical and strategic communications by the use of high quality civilian landlines; what many felt was a fudge that tended to give an unrealistic view of the real military communications needs. An overview of the tactical communications, as deployed in the Gulf, is shown at Figure 3.1.

Strategic Communications

Strategic command and control of Operation GRANBY was carried out at the Joint Headquarters (JHQ) at High Wycombe in the UK.

There was a need for strategic communications links into Headquarters British Forces Middle East (HQ BFME) at Riyadh, in Saudi Arabia, as well as to each of the operational air bases used in the Gulf, and the main port of entry at Al Jubayl. These links were provided by a variety of satellite communications (SATCOM) terminals using the three *Skynet 4* range of UK satellites and operating into an earth station in the UK. The satellite links

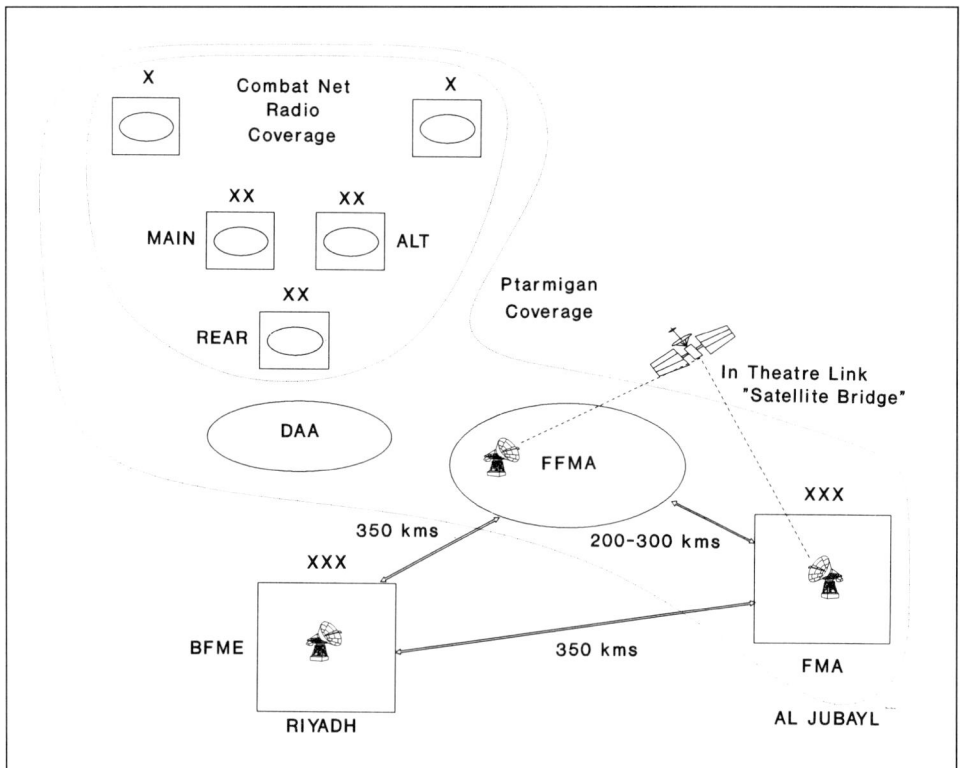

FIG 3.1 Overview of Tactical Communications

were supplemented wherever possible by international landlines and were backed up by long haul HF radio operating into the Defence Communications Network (DCN).

New secure speech systems, known as PATRON, MENTOR and ADVO-CATE were developed for use by planning and operational staffs throughout the UK, BAOR and the Gulf. In addition, new local distribution networks were set up, based on the EUROMUX system, for use in Riyadh, Tabuk and Dhahran. An overview of the strategic satellite communications links is shown at Figure 3.2.

Information Systems

Tactical Systems

The main tactical information system used within the Division and by the logistic staffs in the Force Maintenance Area (FMA) was DICS, the Desert Interim CIS System. This was specially developed for the operation and, in

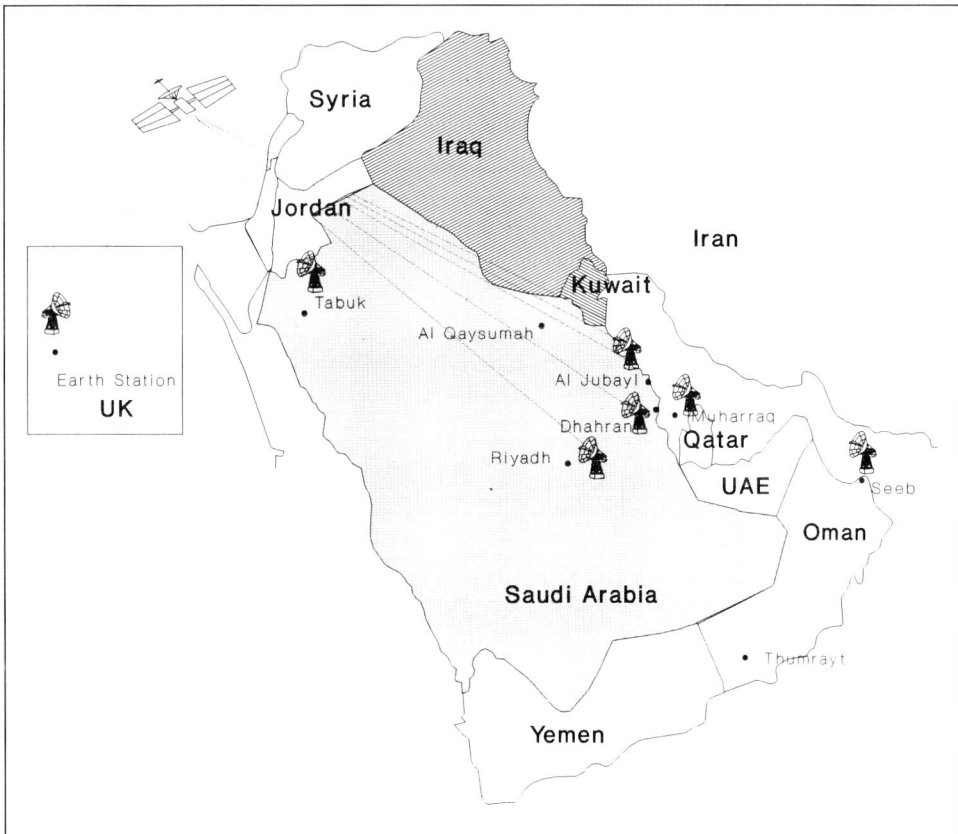

FIG 3.2 Strategic Satellite Communications

part, acted as a substitute for *Wavell*. It was carried over *Ptarmigan* and the satellite links. Other systems used at the tactical level included BRUHN for NBC work and the US systems GOLDWING and T-VITS for intelligence.

Strategic Systems

At the strategic level, the Air Staff Management Aid (ASMA) was the primary information system used for command and control and this all informed tote facility was widely distributed over the satellite links. The Command Support System (CSS) MAPPER was used primarily for casualty reporting and this too was carried by satellite links. A Field Records system was also developed, mainly for use within the theatre. Key systems in use for logistic purposes included the current in-service systems of OLIVER and COFFER. Superimposed on the OLIVER system and operating from the computer centre at Bicester was a classified Log Email facility that was found to be of particular benefit to the logistic staffs. Initially OLIVER was carried by landline but, because of its strategic importance, it was subsequently switched to a satellite link.

A summary of the principal information systems is given in Figure 3.3. Altogether, there were some 25 different systems that were used during the operation.

Development of the CIS Facilities

This chapter has provided a brief introductory overview of the CIS systems that were eventually used in the Gulf. Few facilities were available at the beginning of the deployment and why these particular ones were selected and how they were developed is the principal theme of the next three chapters.

- **DICS - Combat Support and Logistics**

- **BRUHN - Nuclear Biological Chemical**

- **ASMA - Strategic Command and Control**

- **Mapper - Casualty Reporting**

- **Field Records - Personnel and Casualty Reporting**

- **OLIVER - Logistics**

- **Log Email - Logistics**

FIG 3.3 Principal Information Systems

4.

How We Got There: I – Build-up

Nothing is more expensive than a start.
                                    Friedrich Nietzsche, 1888

Introduction

For Operation GRANBY, a most complex web of CIS systems had to be established by those involved, literally as they went along. Furthermore, planning assumptions had to be modified repeatedly as the size and location of British forces in the Gulf changed. As has already been described, few facilities existed prior to this operation that were committed or available for OOA use. New communications systems were developed and existing ones modified whilst work progressed on developing many new information systems specific to the needs of this operation. All of this was done in a very short period whilst the deployments of men and equipment to the Gulf were actually taking place.

That the CIS aspects of the operation were as successful as they were is a great tribute to the many people involved and to their skills and adaptability in constructing effective and much needed systems from limited facilities at such short notice.

Experiences of CIS Staff

A point which we made in our earlier book and to which we return on a number of occasions in this, is that the boundary of CIS includes people. That is, the people responsible for specifying, developing and using CIS are as much a part of the system as the computer hardware and software. The eventual shape and characteristics of the systems depend upon a whole set of complex interactions between people and their perceptions of the current state, of the future requirement, and of the priorities to be observed.

The next part of this book looks at the process of evolution of the major battlefield CIS used in the Gulf and attempts to identify the more important human influences and interactions between them. The story is essentially that of some of the people involved in developing the CIS systems for

GRANBY and of the problems they experienced and had to overcome. From these experiences we should be able to abstract clues as to how future CIS strategies, organisations and facilities might be made ready such as to ensure that we are much better placed for the next time.

Structure of the Chapters

The next three chapters aim to follow the CIS developments through a broadly chronological sequence of the build-up to the operation, through the air war and into the ground war, though there are some deliberate 'flash backs' as well. The experiences related here have all been drawn from detailed recorded interviews with many of those who were directly involved in the CIS activities for GRANBY, and had firsthand knowledge of the events which took place. In presenting this aspect of the story we are faced with some difficulties. The views expressed are not official; they are personal to the individuals concerned and are a set that has been selected and edited by the authors. Whilst we have attempted to carry out this process objectively and without bias, some valid viewpoints are certain to have been missed and others may well have been glossed over or over-emphasised. These difficulties notwithstanding, we believe that what follows does provide as accurate an insight into the CIS activities of Operation GRANBY as it is possible to achieve with an after-the-event account. All the incidents described can be traced back either to firsthand accounts, to official reports or to open publications. Where we have received conflicting views we have tried to represent both sides equally.

Project DICS

The Naming of the Project

DICS is an acronym for the Desert Interim CIS System (though in some official reports this is alternatively shown as the Desert Interim Computer System) and this name had evolved from something that was then being called HICS: the Headquarters Interim CIS System. HICS was planned to be a small Unix-based computer facility designed for the 1st British Corps. It was intended that the HICS system should initially be used in barracks, with the understanding that it might, at some later stage, be deployed to the field. However, in the summer of 1990, its procurement had only just begun.

In early 1990 a Headquarters CIS Group was established under the auspices of the Director of CIS (Army). Its role complemented that of the Arms CIS Group, which had been established at Bovington Camp some two years earlier, tasked with the elaboration of the concept of a Battlefield Information and Control System (BICS) to provide CIS support at battlegroup and below. The HQ CIS Group's task was to look at the concept for a formation level manoeuvre control system which would ultimately replace *Wavell*.

The Group was headed by a full colonel, late of the infantry, who had been introduced to the concepts and practicalities of battlefield CIS support during a tour of duty as an exchange officer with the US 9th Infantry Division (9 ID) in the United States. 9 ID is a light infantry division, used by the US Army as a test bed for CIS and other 'high-tech' developments. Their approach was one of evolutionary prototyping, with the user as developer. By giving users a few basic off-the-shelf hardware and software tools which they were allowed to adapt, the system 'grew' from the bottom up, or rather, from inside the users' organisation outwards. In that way it was intended to integrate itself into the user's way of working.

The 9 ID approach of 'user as developer' had certainly proved itself to be valuable for the testing of concepts, but it was also found to have some serious drawbacks. The British Army lieutenant colonel who was later responsible for the initial testing of DICS had, during an earlier attachment to the US Army, also been responsible for the testing of some of the 9 ID systems in the United States. He reports that this approach often failed to provide systems which could be fielded to the Army and be supported in combat situations. The software was frequently undocumented and contractors had to follow the units round in civilian vehicles in order to fix the programs as they failed. Failures could be induced by the most primitive stress conditions. As a result of this experience, the US Army subsequently modified the 9 ID activity such that it now only tests concepts.

A similar but rather more restricted approach was adopted in the UK by the 3rd Armoured Division, which had been using laptop computers in support of staff work both in barracks and in the field for about two years. As we shall see, this experience of a novel and apparently highly successful way of introducing and developing CIS proved to be a key driving force for the DICS project. Nevertheless, the warning indicators from the US Army experience would need to be noted carefully.

The HQ CIS Group was staffed by officers with operational staff experience, two of whom happened to be professionally qualified engineers. It was based at Feltham, co-located with the Directorate General of Military Survey. That co-location was significant, and recognised the vital role to be played by terrain databases and the geographical information systems (GIS) concept in any future manoeuvre control system. Events were to provide an early opportunity to try out the underlying principles behind what eventually became known as the 'Almonds Strategy'.[1]

The Iraqi invasion of Kuwait in August 1990 and the likelihood of the committal of UK land forces in response presented the ideal opportunity for the HQ CIS Group to do some evolutionary prototyping of their own. The original concept was to advance the procurement of a ruggedised HICS as an Urgent Operational Requirement (UOR). At the time the HICS architecture consisted of a number of Unix-based server systems with perhaps up to a

---

1 This strategy document took the name of the chairman of the working party responsible for its formulation.

dozen terminals attached to each. Following discussions between the HQ CIS Group, Corps Headquarters in Bielefeld and HQ 7th Armoured Brigade at Soltau, an initial DICS bill for 7th Armoured Brigade of some 130 terminals was arrived at.

Finding a Champion

Thus Desert HICS or DHICS was born; and this soon became renamed DICS as the silent 'H' got dropped. However, the concept of such a project certainly did not meet with universal approval. There were those who referred to DICS as the 'Urgent Operational Solution' – a solution looking for a problem. The project might well have died an early death but for the robust enthusiasm, drive and determination of its main protagonist, aided and abetted by the CIS Staff at Corps Headquarters and the supportive insistence of the Deputy Commander of 7th Armoured Brigade that some such system was going to be essential in the coming operation. What was clear was that there were no contingencies for any of this in terms of either staff or equipment and someone was going to have to grasp the nettle; that someone turned out to be the Head of the HQ CIS Group.

This is not to suggest that the Head of HQ CIS Group was the sole driving force; he would be the first to deny that and most vehemently. There were many key players without whom the project could not have succeeded. However, the role played by the Head of HQ CIS Group does provide a good example of the case that if you want something done in a hurry, you should identify a 'champion' with the necessary vision, drive, resources and cross-organisational authority and let him get on with it. Of course, he must have the support of many others, but he provides the focus for getting the job done. This too has it drawbacks as we shall see. In the headlong dash to try to get the job done, essential moderating management controls may be addressed either inadequately or simply ignored and overridden. User trials, configuration control and quality assurance procedures may be viewed by some as little more than bureaucratic nuisances, which hinder the process of getting the products out to the field as quickly as possible. This is especially true if some other agency is made responsible for these aspects.

We shall return to that issue. Meanwhile, the reader, having noted that *Wavell* formed part of our 'ideal' CIS configuration in Chapter 2, should by now be wondering why the BAOR units had not planned to take this existing system with them. After all, *Wavell* was an immediately available in-service system with which they should all be familiar. Why was there any need at all for a Desert Interim CIS System and its associated UOR? Why not take *Wavell*?

Problems with Wavell

Perceptions, particularly after the event, vary. Some note that, initially, only 7th Armoured Brigade was earmarked for Operation GRANBY and, as is still the case, there was no real role for *Wavell* at the brigade level. In any case, *Wavell* was perceived as providing only very limited support for logistics and this was rightly thought to be one of the key areas that would be requiring computer assistance.

There was, however, another good reason. One senior commander who had been exposed to the now famous GRiD trials that had been run in the 3rd Armoured Division was quite clear about his requirements: 'I want laptop computers; I want GRiD computers'. He knew from firsthand experience how useful and effective the ruggedised GRiD laptop computers could be in an operational environment.

Support for the Brigade

With *Wavell* out of the running, PC-type laptop computers, and preferably GRiDs, were placed high on the wish-list of the staff for their integral information systems. The question was, what equipment could be procured in the short time available? This problem was addressed in Germany by the Corps HQ CIS staff, assisted by the head of the CORMIS (CORps headquarters Management Information System) team, who was about to leave Corps Headquarters to take up a new post in the Army CIS Agency (ACISA) at Blandford.

There was a strong feeling that having to deploy a brigade on its own would be very difficult indeed from both a CIS and a logistic viewpoint. The brigade was designed to operate in a BAOR scenario with division and corps support environments to back it up; but, operating offensively over extended distances and away from this latter support, it would require additional CIS and logistic support. What IT support could be provided in the time and what would be the most useful and effective?

The CORMIS view was very much against a GRiD hardware solution. Major problems were foreseen over the security of disks, with networked systems and the lack of expansion facilities. There was also concern about the brigade being offered what were somewhat disparagingly referred to as 'rubber crutches' which would fall over as soon as they were put under stress. MSDOS, the operating system used by most PCs and by the GRiDs, was not a good basis for a resilient systems architecture and some Unix-based facility would be much more preferable. Other voices, however, were strongly supporting the use of GRiDs, not so much from positions of systems understanding or expertise as from the more pragmatic viewpoint of availability and the fact that many of the staff liked GRiDs and had experience with their use. In the end, the pro-GRiD lobby won, at least in BAOR; but, that still left the problem of software.

The Influence of CORMIS

At this time the CORMIS project was in its second phase. The CORMIS team had just finished upgrading the hardware installations to some 90 terminals and six Motorola 68030 processors which were running under the

Unix operating system. Word-processor and spreadsheet facilities were being provided by means of the UNIPLEX package and INFORMIX was being used for database facilities. This would all have formed an excellent basis for the DICS software. Many staff were already familiar with it and it covered the majority of the required functions. The problem was, of course, CORMIS was all Unix based and the GRiD solution would use MSDOS.

As with so many aspects of this operation, luck played a crucial part once more. A few months previously, some of that CORMIS software had actually been ported on to MSDOS based systems. This was not part of some grand plan, which had somehow rightly predicted a future operational need. What had happened was much more mundane. The maintenance contract for the rather elderly CPT word processors, then being used by the typing staff throughout 1st British Corps, had run out and the systems had become obsolete. DGITS IT(OS), the branch responsible for typewriter support, decided in their wisdom to replace all the CPT systems with Wang PCs and Wang's own word-processing facility. This proposal was met with some horror by those who would have to become proficient in yet another word-processing package. The CORMIS team was therefore prevailed upon to port some of the CORMIS software, and in particular the by now familiar word-processor, from Unix-based CORMIS to DOS-based Wang using UNIPLEX.

There was thus some recent expertise in the CORMIS team in converting from Unix to MSDOS and they had some software already available. This talent was harnessed by the Head of the HQ CIS Group, who despatched his two staff members from Feltham to Bielefeld to establish a DICS development cell there, and eventually bring it back to the UK.

Both of the officers sent to Bielefeld had considerable practical engineering experience. The SO1 was one of only two serving officers to have attended both the Army Staff Course and the RE Long Engineering Course. He quickly identified the component products of DICS, defined milestones and applied PERT techniques with Harvard Total Project Manager software. He never missed a milestone. Here too was one of the key players without whom the DICS project might never have succeeded. The SO2 was trained as a REME engineer and had already applied his skills to eight other software-intensive projects. His role was to insist that all deliverables were formally signed off by users and technical Quality Assurance (QA) members of the team, that documentation was developed alongside the software and that configuration control was applied from the outset.

It is most interesting to note that at this stage the co-ordination of DICS, including the 'redirection' of personnel to work on the project was almost entirely *ad hoc* and unofficial. The Head of the HQ CIS Group sought to recruit any individuals with the right knowledge and skills on to the DICS Development Team. One of the people who slipped through his net was the ex-CORMIS team leader who, he felt, would have been an ideal addition to his organisation. However, as we indicated earlier, the ex-CORMIS team leader had been posted from Bielefeld to ACISA at Blandford; there he was employed as the head of a separate DICS support team under the control of

Head ACISA. We consider in a subsequent section the difficulties and tensions that arose from having initially completely separate development and support organisations for DICS.

As the DICS project became established, albeit in this somewhat *ad hoc* manner, so the development and support organisations became formalised and an Op GRANBY CIS Steering Committee and Project Board were set up under the Director CIS (Army). The position of Head of the HQ CIS Group's role in the project was also formalised, and he added the title of 'DICS Project Coordinator' to his ever-growing and already impressive row of responsibilities.

The DICS User Requirement

One of the first tasks for the newly-formed DICS Development Team was to get a reasonable idea of the user requirement for DICS. They cleared with the Deputy Commander of 7th Armoured Brigade the needs and priorities of the Op GRANBY force, still at this stage of brigade group size. Again, by great good fortune, the Deputy Commander had been Colonel G3 O&D and CIS at Bielefeld; he was both thoroughly familiar with the CORMIS project and he placed G1/G4 well at the top of the priority list of user requirements for the Brigade.

The priorities for DICS, as originally proposed by the Deputy Commander of 7th Armoured Brigade and subsequently endorsed by the Chief of Staff of 1st Armoured Division when the Op GRANBY force was expanded to a divisional deployment, were agreed as shown in Figure 4.1. Work could now be

---

- **In Theatre support for G1/G4. This should provide for stock control and asset tracking to support the management of logistic assets in the Brigade and Force Maintenance Areas. The system should provide information to the commander on the sustainability of the Force.**

- **The provision as rapidly as possible to Brigade, Divisional, Divisional Rear and Force Maintenance Area Headquarters of the logistic state of Units so that priorities for resupply may be adjusted and implemented.**

- **The development of reports and returns and word processed applications to assist in the production of Operations Orders, Administrative Orders and Movement Orders.**

FIG 4.1 The Staff Priorities for DICS

initiated by the DICS Development Team to produce the software that would aim to fulfil these staff requirements.

Initial Support for G1

The Field Records Unix Boxes

The newly formed Op GRANBY CIS Steering Committee, taking note of hard-won experience from the Falklands conflict, had very sensibly placed an urgent priority on automated support for the Field Records function. The CIS Staff at Bielefeld, who were already discussing with GRiD Computers the purchase of the initial tranche of DICS systems for the brigade, were tasked with the immediate purchase of suitable computers for Field Records. Over one weekend, three tower PCs were obtained from a firm called Victor. These were Intel 80386-based systems with 12 megabytes of random access memory (RAM) and a 600-megabyte hard disk drive, together with an array of tape streamers.

At the same time as the Unix 'boxes' were being procured by the Bielefeld CIS Staff; an officer, already serving in Germany as the SO 2 (Communications) of 1st Armoured Division in Verden, was selected to be the CIS Staff Officer for 7th Armoured Brigade. He was warned for movement to the Gulf, but his first port of call was Bielefeld to meet his first system – the three Unix Boxes, which he subsequently accompanied to the Gulf with the Advance Party of 7th Armoured Brigade.

Development of the Database

A Field Records application was to be written by a programmer in 1st British Corps using the INFORMIX 4th Generation language (4GL) database system as supplied with CORMIS, and a unit paymaster who had computer experience was appointed to be the Officer in Charge (OIC) of Field Records. The provision of the database was no easy matter. It was hoped that full details of all personnel going to the Gulf could be imported into the Field Records INFORMIX database prior to leaving BAOR from the nominal rolls held in the Pay and Records computers at Worthy Down. Although details such as name, number, rank and date of birth could be imported quite easily, one of the most important items from a casualty reporting viewpoint could not: that of next-of-kin. In the end, something like 6,000 partial records were loaded into the system before it departed for the Gulf.

The Setting up of Field Records

There were no real terms of reference for anyone at this stage. Whatever was done was done because it seemed the most sensible at the time. The newly-appointed CIS Staff Officer of 7th Armoured Brigade flew out with the Advance Party for 7th Armoured Brigade because that was when the Field

Records computers were being sent out; he based himself, initially, in the Force Maintenance Area at Al Jubayl because that was where the Field Records organisation was to be established.

Al Jubayl is a port some 20 kilometres from the airport through which all in-bound troop movements were being routed. Field Records was set up in 'Shed 4' at the port itself and all incoming personnel were driven from the airport to Al Jubayl and processed through this reception centre. Here they received indoctrination briefs on such matters as the *Scud* threat and the current NBC state and then they were required to report their personal details to Field Records.

The Field Records Architecture

The Field Records architecture was that originally assembled in Bielefeld. It consisted of two of the three 80386 tower systems connected together by an Ethernet cable in air-conditioned offices along one end of Shed 4. One of the computers was made the operational system and the other a hot standby. It was designed such that approximately every 15 minutes a standard Unix process, known as cron, would copy the database, in its entirety, from the operational system to the hot standby system. This ensured that, in the event of a major failure, the database would never be more than 15 minutes out of date. The third processor system was installed physically remote from these two. Backups were taken, twice daily, from the standby system so as not to slow down the operational system, and these were used to maintain a full fallback capability on the third, physically remote, system.

Initial Field Records Procedures

Serial cables were laid from the operational computer down to some trestle tables positioned in the entrance to Shed 4. Six Wyse 60 dumb terminals were set up on these and behind them sat the Field Records clerks. Everyone in-bound to the Gulf had to file past one of these terminal positions and confirm his personal details with a clerk. For those personnel coming from the major units in BAOR this was not too much of a problem since the majority of their details were already in the system, other than their next-of-kin. For some of the UK-based units, however, such as 33 Field Hospital which had been brought in from all over the United Kingdom, there were no pre-existing records in the database and the clerks then had to enter in the full details of everyone in the unit.

Once it was believed that the last member of a unit had been through this processing, a full listing of all the personnel details then held by Field Records would be given to the Chief Clerk or Adjutant of the unit. This was so that a check could be made against the unit's own mobilisation details; these were printouts that it was hoped had been brought with them from their peacetime PAMPAS systems. By a process of double checking and re-interviewing as necessary, it was hoped that all of the personnel records problems could thus eventually be ironed out.

There was a total of eight clerks in Field Records at this time and a team of six was needed, working in shifts 24 hours a day, to man the terminals whenever a flight came through. Watching the six clerks struggle under the sheer weight of numbers, the system's developers were under no illusion as to the inability of the procedures to cope if a force larger than the Brigade was ever to be processed. Whilst these *ad hoc* arrangements were barely acceptable for the initial stages of the operation, some real improvements were going to have to be made if the full Division became committed.

There Must be a Better Way

Why, they thought, should not all of this checking and updating have been done ahead of time on the units' own PAMPAS systems in their in-barracks peacetime locations? Given that, the details could then simply be imported on to the Field Records system by means of a single floppy disk brought into theatre by the unit. The PAMPAS team at Worthy Down, however, rightly pointed out that the PAMPAS computers of the units were completely incompatible with the Field Records computers: PAMPAS made use of the BOS Operating System running on a Pertec; Field Records were using the Unix operating system running on an 80386; the associated disk formats are very different and neither system can easily be persuaded to read the disks of the other.

The solution devised by the Field Records team at Al Jubayl was simple but effective. Having ascertained from Worthy Down that the PAMPAS computer could be made to print personnel details in the format required by their Field Records system, they suggested the replacement of the serial printer by an MSDOS-based personal computer, connected via its serial port. Then, instead of the stream of records being printed on to paper, they could be directed straight to a file on the PC's hard disk. That file, which would be in MSDOS format, could then quite simply be converted into a form acceptable to the Unix Field Records system.

Despite some initial reservations, development of such a facility became a matter of real urgency when word came that the 1st Armoured Division was to deploy; Field Records would otherwise simply not be able to cope. Worthy Down were persuaded to send out to the Gulf a PAMPAS Pertec machine together with the terminal emulator and editing software. Meanwhile, tests were carried out at Worthy Down to prove the concept, which was subsequently shown to work well, and then revised software was written for PAMPAS to produce a new printed report that contained just the information required by Field Records.

The New Procedures

The idea now was that all the Army units in the UK and BAOR which were earmarked for the Gulf should be instructed to get their PAMPAS records fully up to date. When that had been done, clerks would be despatched to each unit concerned and they would collect a complete copy of the unit's PAMPAS records on floppy disks. These disks would then be flown out to the Gulf. There, using the Pertec machine in an off-line mode together with the new PAMPAS report software, a PC, the Crosstalk emulator software and a WordPerfect editor, files would be generated that were suitable for the Unix Field Records system and which contained all the personnel details of the units concerned, in precisely the form required by Field Records.

An additional field was included in each personnel record indicating 'checked' or 'not-checked'. The procedure adopted was for the clerks at the terminals in the Field Records reception area to check the personal details from the floppy disk files with each individual coming through, and then record that this check had been carried out. Checked records could then be added to the live Field Records system with a high degree of certainty that they were now correct. The procedure thus became one of checking the unit's PAMPAS data with the individuals concerned as they arrived in theatre rather than the much more time consuming and error-prone process of entering new data at the terminals.

DICS and the Brigade

Meanwhile, the first tranche of DICS that had originally been bought by the CIS staff at Bielefeld for training purposes in BAOR had been taken by 7th Armoured Brigade with them to the Gulf. These were GRiD 1530 Non-TEMPEST laptops with four mbytes of RAM and 40-mbyte hard disks. The DICS Development Team had configured these systems with a number of software packages whilst they were still in BAOR and had put Army Formation SOPs on to them, but the equipments were not intended for operational use because of the TEMPEST factor. They did, however, give an indication of the kind of software that it was intended would be available on the later operational equipments.

The Need for Computer-to-Computer Communications

Having established himself in-theatre and got the Field Records work under way, the 7th Armoured Brigade CIS staff officer turned his attention to DICS. He appreciated the value of the work of the DICS team in Bielefeld, but was concerned that they had so far not implemented a most important facility: that of computer-to-computer communications over *Ptarmigan*. The development team had, as it turned out, good reason for this: the facility had yet to be developed and submitted for trial before the concept could be incorporated into DICS. The CIS staff officer however decided to investigate the provision of limited facilities in-theatre himself. The simplest approach was to load each of the computers that were to be connected together with a commercial communications package; the one chosen was called Crosstalk – the same general-purpose PC communications package that had been used in the Field Records PAMPAS-Unix conversion described earlier.

By setting up one of the computers to run Crosstalk in receive mode, users could dial in to this computer from any other machine which had the communications package and transfer data on to the hard disk or floppy disks of the receive computer as required. The Crosstalk package would, in addition, allow the remote user to command the receiving computer to change disk directories and so forth.

The First Link

The first such computer-to-computer link over *Ptarmigan* was set up between the FMA at Al Jubayl and HQ 7th Armoured Brigade, when the headquarters first deployed to the field; this was done because of the large amount of paperwork that needed to be sent between the two. It was implemented by means of what were colloquially known as 'WMIS leads' – pronounced 'Wimmis', these having been 'borrowed' from 1st British Corps. The WMIS leads enabled a GRiD computer to be connected to the *Ptarmigan* data adaptor and had originally been developed for the *Wavell* Management Information System (WMIS), hence the name. The WMIS system consisted of a bespoke software package that had been developed to run on GRiD Compass machines for the in-service management of *Wavell* within 1st British Corps.

Using these leads, the GRiD computers were connected to the *Ptarmigan* data adaptors, with the receive system being switched permanently to 'auto-answer' mode. A user simply dialled into the receive data adaptor, switched to data, and immediately got the remote computer prompt back on his machine. He could then transfer files between the computers as he wished. There was, of course, one major problem. The GRiDs in the Gulf at this stage were the non-TEMPEST training machines and so this link could only be used as a demonstrator and not for operational purposes.

DICS Support and the CIS Chain of Command

One of the key difficulties at this stage was the lack of any clear management structure or technical chain of command. Initially, 7th Armoured Brigade G3 CIS Staff (one major and two sergeants who were also the entire DICS Forward Support Team) had responded to tasking from G3 CIS at HQ 1st British Corps in Bielefeld. Once in the Gulf they found themselves, almost by default, answering in part to the DICS team in the UK and subsequently under the loose technical control of ACISA. The only in-theatre CIS point of contact to start with was the Commanding Officer of 30th Signal Regiment, who was based at Riyadh as the senior communications officer (Commander Communications) of the UK Land Force contribution to the Coalition.

As well as the problems created by multiple tasking, such an *ad hoc* chain of command made it difficult to obtain formal MOD support or action to meet the local operational needs. Trying to telephone round the UK and BAOR from Saudi Arabia in order to make contact with the right person was no easy task. For the Field Records effort, however, an excellent point of contact was established with a senior member of the Unit Computing (UNICOM) project team at Worthy Down, who was a strong champion of the Field Records cause and effectively fought the UK Field Records battles on their behalf.

Linking HQ 7th Armoured Brigade to 1 MARDIV

At this time, communications problems were also beginning to arise over the liaison between the 7th Armoured Brigade Headquarters and the 1st Marine Division (1 MARDIV) under whose control the brigade would operate. Initially, the 7th Armoured Brigade Liaison Officer (LO) took *Ptarmigan* Single Channel Radio Access (SCRA) facilities with him to HQ 1 MARDIV. However, because the Marine Division was initially deployed quite forward, this had the effect of extending *Ptarmigan* SCRA coverage well forward of the rest of the trunk network. In addition, the need for a number of different desks in 1 MARDIV to have regular contact with their colleagues in 7th Armoured Brigade suggested that the most sensible option would be to deploy a *Ptarmigan* node at HQ 1 MARDIV. It was also a good opportunity to show *Ptarmigan* off to the Americans! This deployment was done using what in Germany had been the secondary access node from HQ 1 Division Rear.

As soon as the *Ptarmigan* node arrived, the Americans saw this as a good opportunity to connect HQ 7th Armoured Brigade on to their information system. Could it be done over *Ptarmigan*? The 7th Armoured Brigade CIS Staff Officer, having cautioned about the need for a TEMPEST computer, took himself off to HQ 1 MARDIV, once more clutching his WMIS leads. In the Marine Division headquarters he found a G6 organisation for communications and information systems with a major who was the Information Systems Management Officer, or ISMO.

The ISMO was quite happy to lend HQ 7th Armoured Brigade a TEMPEST-approved computer, which turned out to be an older version of the DEUCE (Downsized End User Computing Equipment) system, a number of which were subsequently bought for the DICS project. Local tests on the *Ptarmigan* node with slightly modified WMIS leads eventually proved fruitful, so the next step was to reproduce the set-up at the 7th Armoured Brigade end using the DEUCE and the WMIS leads, and to try the system over *Ptarmigan*. The brigade staff were surprisingly unenthusiastic at this stage; 'Not another computer? What are we supposed to do with this?' were typical of their reactions.

In fact, once the link had been successfully established, views changed quite radically and the system proved to be very useful; especially for items such as Air Tasking Orders which were 80-page signals sent out from

1 MARDIV. Using the computer at HQ 7th Armoured Brigade, the Order could be received and stored on the hard disk and only the parts relevant to the Brigade printed out locally.

A Success that Was Missed?

This early linking of HQ 7th Armoured Brigade to the American network was possibly the first successful instance of allied CIS operating 'in anger' over the *Ptarmigan* system. There was, in addition to the dial-up facility, a US electronic mail (email) system and a CHAT system; the latter permitting two users to communicate informally through split screens. This achievement was formally reported back to the UK, as was the general approach to linking PCs over *Ptarmigan*.

In the end, of course, it all became rather academic; the Brigade did not long remain under control of 1 MARDIV once the 1st Armoured Division started deploying, and then all the links between the brigade and 1 MARDIV were discontinued.

Casualty Reporting

At about this same time, great concern was beginning to be felt over the potential difficulties surrounding effective casualty reporting. With the world spotlight focused on the Gulf and electronic news gathering (ENG) teams able to report instantly, by means of satellites, live television coverage of events as they occurred, there was a real risk of casualties being identified publicly before any official details were available to advise worried next-of-kin.

Estimates of the number of casualties that might occur and the timescale over which this might happen, even for just the 7th Armoured Brigade area, were particularly worrying. The focus for all such notifications had to be Field Records. There were two CIS problems here: how to get casualty reporting information very quickly back to the UK for official release to next-of-kin; and, how to collect and collate such information from within the theatre of operations.

Casualty reporting procedures require the use of the NOTICAS signal. With his communications background, the 7th Armoured Brigade CIS Staff Officer was well aware that the clerical effort required to type up large numbers of NOTICAS signals could quickly swamp the Comcen, run by 30th Signal Regiment, that provided the formal message link back to the United Kingdom. He therefore proposed that the NOTICAS signals be prepared direct by the Field Records system. By means of a Trend ESR 615 teleprinter and a paper tape punch, the five unit code tape required by the telegraph system could be generated direct from the Field Records computer. This had the additional advantage that the personal details would be automatically validated; they would have come direct from the Field Records database, the records of which had been checked with the individuals concerned and they would not have been, perhaps incorrectly, retyped.

Why Not Use Mapper?

At about this same time, proposals were being made in the United Kingdom that Mapper, the UKLF command and control system, should be used for casualty reporting. It was decided that Mapper facilities should be put into 33 Field Hospital located in the Goodyear factory in Al Jubayl, into Field Records itself and into 22 Field Hospital in Bahrein.

However, 'What is it for? How is it to be used?'; these were questions to which the in-theatre CIS staff, including OiC Field Records, felt that they were receiving no clear answers. The fundamental issue here is that all casualty reporting out-of-theatre has to come from one authority: and that authority is Field Records, where the personnel records of all those in theatre reside and with whom the responsibility rests to validate any casualty details. The in-theatre staff felt that any suggestion for the UK to have more advanced casualty information direct from the field hospitals via Mapper had to be discouraged as a thoroughly unsound practice.

On the other hand, the imperative in the UK was the need to inform next-of-kin before they learned of death or injury through the media. Nobody really disagreed; Field Records had to be the focal point. The eventual compromise was that Field Records would have four terminals on the Mapper system and when the clerks received and typed up NOTICAS signals, they would 'rotate their chairs and type the information into Mapper as well.' (Readers of our first book, or students of NATO interoperability will recognise this as an example of 'Degree 2 Interoperability'). That way the information from Field Records would get straight back to the responsible MOD Branch (PS4 Cas) who would also be equipped with Mapper terminals.

What about Communications?

The first problem to be solved was that of providing the necessary strategic communication links between the UKLF Mapper computer system, based at Wilton, and the proposed terminals in the Field Hospitals and at Field Records in the Gulf. Also to be borne in mind was the likely need for tactical communication links when Field Records, and perhaps elements of the Field Hospitals themselves, had to deploy further forward.

The suggestion put forward by the JCIS staff at Joint Headquarters in High Wycombe was for the strategic links to use landline via the local PTT. The result of this was that shortly thereafter Saudi Telecom engineers were found putting into the CIS office in Shed 4 at Al Jubayl a private wire, linking direct to Wilton, for the Field Records Mapper terminals. However, it took several days, after a suitable modem had been acquired from 30th Signal Regiment, which they had previously used for ASMA, to determine by trial and error, line and modem settings that would permit effective intercommunication with Wilton. Although this approach did eventually work and gave an unexpectedly good 9.6 kbps synchronous transmission rate, there was great concern at the Gulf end of the circuit about its sustainability.

Local CIS staff had been warned that Saudi Telecom could prove more vulnerable than a military system in the event of hostilities. They were only too well aware that this Mapper private wire circuit was patched some 12 times between Al Jubayl and Jeddah and, in fact, patched three times in Al Jubayl itself, was carried over many different bearers, was frequently switched between digital and analogue forms, and had an unacceptably large number of balancing transformers along its length. The slightest stress and the circuit would rapidly become unworkable.

It would be possible, of course, to use satellite communications. At this time, there was a single TSC 502 satellite terminal at HQ FMA in Al Jubayl, which was where Field Records was based and this could be configured to carry the 2.4 kilobits per second Mapper circuit for Field Records use. However, there were no satellite links planned to go into the Field Hospitals so Satcom was not an option for them. It was for this reason, then, that the landlines were put in to the hospitals to carry the Mapper circuits.

Employment of Mapper

How was Mapper to be used? The idea, referred to previously, that the clerks should 'rotate their chairs' and retype the casualty reporting information into the Mapper terminals after having updated the Field Records system, did not gain much support at Al Jubayl, for rather obvious reasons. There was just too much to do at Field Records without the time-wasting task of typing the same information in twice and, besides, there was a significantly increased risk of errors in using this approach. What was really required was some automated link at Al Jubayl between the Mapper and Field Records systems. As with so many of the CIS issues that arose in the Gulf, this was yet one more instance of an interoperability issue, or more properly, the lack of it. The Mapper system uses non-standard, proprietary synchronous terminals, so an ordinary PC could not be used direct as a terminal on to Mapper equipment. Such an approach would have been the simplest, since MSDOS files could then have been taken off Field Records, loaded into the PC being used as a Mapper terminal and then down loaded directly into the system. However, after some investigation, a plug-in printed circuit board (PCB) card and a software package for the PC that would make it act as a Mapper terminal were located. With these facilities flown out from the UK, the clerks were then able to update Mapper automatically with casualty information which was taken direct from the Field Records database.

How could Mapper most usefully be employed at the hospitals? What was really required here was a means of updating Field Records with casualty information from the hospitals; this would then be passed formally from Field Records, over Mapper, to notify MOD in the UK. However, the Mapper system could also be used to link the hospitals with Field Records. To this end the latest version of the Field Records database was downloaded into Mapper and a simple retrieval system was written for Mapper that allowed the hospitals to access personal details from this copy of the Field Records database. This facility was used by the hospitals for their Admissions and Discharge Books. The hospitals were thus validating their entries against the latest known information at Field Records and casualty information at the hospitals was also available to Field Records at Al Jubayl. Subsequently, a similar Field Records database retrieval facility was developed to run on DICS as a backup for the Mapper capability.

An Admissions Exercise

Before the validation facility was available, an admissions exercise had been run in Al Jubayl using 33 Field Hospital and the American Marines hospital. The scenario was a *Scud* landing on the tented city in Al Jubayl, and the exercise was to practice the casualty procedures. After the event, the admission and discharge records of 33 Field Hospital were compared with the Field Records information and it was found that almost a third of the entries could not be made to tally. Although, in many cases, it was possible to second guess to whom the entry might refer, OiC Field Records could not accept such a level of inaccuracy for casualty reporting purposes. Had the exercise been reality he would have had to refer all the doubtful entries back to the hospital for confirmation. The database validation facility was thus seen to be a vital element in achieving an acceptable standard of casualty reporting.

A Focal Point for Casualty Information

The next issue that had to be considered was how Field Records would need to be structured and organised in order to cope with a full divisional deployment. Where was the most likely focal point for casualty information going to be? For the ground war, one likely candidate was thought to be at the Headquarters of the Divisional Administrative Area (HQ DAA) if, that is, it was decided to deploy an HQ DAA. Should Field Records be deployed there? Two strong reasons against could immediately be identified: it was unlikely that there would be strategic communications back to the UK from HQ DAA, and it was rather far forward to risk placing the whole of the Field Records facility for the theatre. However, it might make good sense to place a forward reporting cell of Field Records at HQ DAA. The only problem then would be the provision of appropriate data communications between terminals at the forward reporting cell and the Field Records system in Al Jubayl. Why not use *Ptarmigan*?

Earlier work in setting up a data link between the Headquarters of 7th Armoured Brigade and 1 MARDIV had shown the feasibility of using *Ptarmigan* for this purpose. The problem, however, was that one terminal link was going to be insufficient; at least four were going to be needed and only one 2.4 kbps *Ptarmigan* circuit could be spared. The requirement was referred to Joint Headquarters (JHQ) back in High Wycombe who were responsible for all communications and CIS planning. The case was made as to why four terminals would be needed at the forward reporting cell and why these could not be connected to a local Field Records processor with a single link back to the main Field Records processor at Al Jubayl; this latter approach being ruled out because of all the problems of distributed database working for which there was neither the expertise nor the software available.

There were differing views as to the feasibility of providing a technical solution to this problem. The issue hinged on the feasibility of using statistical multiplexers. By utilising the 2.4 kbps *Ptarmigan* speech circuit as a 16 kbps synchronous circuit connected to a statistical multiplexer at each end, it was possible that four 9.6 kbps asynchronous data circuits could be connected to the multiplexers. Although the four 9.6 kbps circuits would aggregate to well over double the channel capacity of 16 kbps, the system works because the probability of all four circuits requiring channel capacity at the same time is small. Once it was demonstrated that this solution was viable, additional links between Field Records and the hospitals were proposed with the use of Single Channel Radio Access (SCRA) over the *Ptarmigan* system. Because SCRA was, in any case, being put into the hospitals and more clerks for Field Records were being made available, it seemed sensible to set up a clerk with a terminal in a small Field Records cell at each hospital who could then access the Field Records computer over SCRA. This would provide a back up to the Mapper facilities.

An Implementation of the Systems

This solution was implemented in the second half of January 1991 in time for the ground campaign, with the use of the statistical multiplexers over *Ptarmigan*, with four terminals forward at HQ DAA working at 9.6 kbps. The system was found, in practice, to work very well. Also set up at this stage, and found to be effective, were the terminals at the two Field Hospitals working at 2.4 kbps over the *Ptarmigan* SCRA system.

The last contribution made by the now ex-CIS Staff Officer of 7th Armoured Brigade before he was allowed to get on with his studies at Shrivenham was in response to an urgent call from the CIS staff at JHQ, seeking his advice on connecting PS4 in the MOD direct to the Field Records system at Al Jubayl, because the Mapper system had crashed at Wilton. Simple dial-up links and PCs running the modified terminal emulator software from the Gulf were implemented in a very short time and this quickly provided a UK backup for the Field Records use of the Mapper system.

The Development of the DICS System

Meanwhile, as the Field Records issues were being tackled and solved by the in-theatre staff, development of the DICS concepts were going ahead in the UK and BAOR under the guidance of the Project Co-ordinator. His two staff from the HQ CIS Group, working in Bielefeld with what had been the

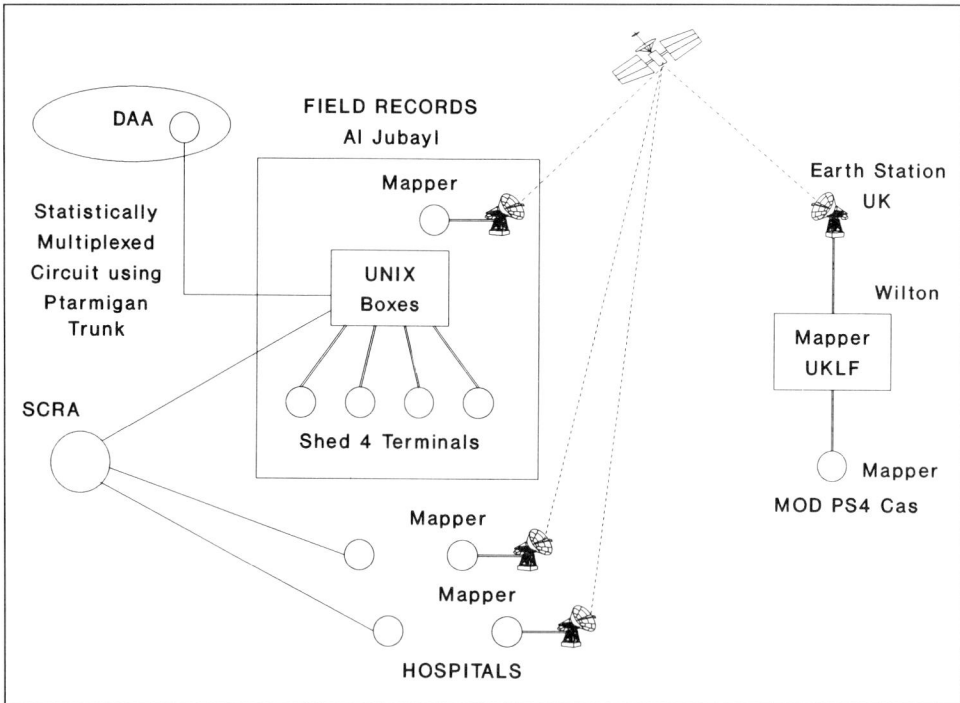

FIG 4.2 Summary of Field Records Systems

CORMIS team, but was now effectively the DICS Development Team, had by this stage obtained the agreed User Requirement for the project from the Deputy Commander of 7th Armoured Brigade. Thinking was now concentrated on a suitable architecture.

An Architecture for DICS

In concept, the team decided to develop DICS by stages and initially supply a number of stand-alone workstations so that anyone who had to manage resources or had to carry out written staff work would have access to a computer to make him more effective. These systems were to provide standard office automation (OA) facilities, such as word-processing, spreadsheets, flat file database access and simple graphics. It was also thought that a more complex database would eventually be needed and so an implementation of INFORMIX as the Relational DataBase Management System (RDBMS) was planned.

It was noted that much staff work using such OA facilities was already being done on privately owned PCs as well as those provided through official sources such as the small systems groups (SSGs) and the Presidents of the Regimental Institutes (PRIs). In the main, these systems were MSDOS-based PCs running standard commercial packages and, if the benefits of these staff experiences were to be usefully employed, it was felt that the DICS architecture should also be MSDOS based and should implement familiar commercial packages.

Whilst it was acknowledged that MSDOS was not the most suitable of operating systems for an operational environment, any attempt to move to a more appropriate operating system, such as Unix, would involve system management functions that could not be supported in the Gulf and a loss of familiarity for the users. However, it was also felt that, in order to survive in the desert environment of the Gulf, ruggedised systems would be needed, and where these were processing classified information or were attached to operational communications systems, TEMPEST-approved machines would be essential. The users' wish to have a small, lightweight computer that could easily be operated in confined and highly utilised areas was also noted. This requirement suggested that the portable, briefcase-size system or lap-top, which could be carried under the arm, was perhaps the most appropriate of currently available PCs.

Having identified a template for the stand-alone systems, the team next considered the means of interconnection. It had already been demonstrated in the Gulf and Germany how two systems could intercommunicate over the *Ptarmigan* circuit switched system using the data adaptors and this had been confirmed formally at RSRE Malvern. This facility was then incorporated into the DICS architecture with the KERMIT public-domain communications package. KERMIT was chosen rather than the Crosstalk package that had already been tested in the Gulf mainly because the source code for it was readily available and it was designed to be used with the printable character set as opposed to a binary stream; this latter issue being of importance in trying to avoid inadvertent code signals being given to the *Ptarmigan* switches by message patterns in the data stream.

The final step in fulfilling the DICS architecture was the implementation of Local Area Networks (LANs) in the larger headquarters environments. DICS was designed, in that situation, to provide a single workstation server, connected through the data adaptor to *Ptarmigan* and via optical fibre cables to a maximum of five local terminal workstations. It was also planned that the DICS server workstation would be sited alongside the RAF Air Staff Management Aid (ASMA) terminal. The suggestion that the two might be electronically linked was considered by some to set a dangerous precedent in that it might be inviting a bypass of the normal chain of command.

The Choice of Hardware

All the analysis to this point had indicated the need for small, portable, TEMPEST-approved, laptop type, MSDOS-based, personal computers and of these GRiDs were already well known and well liked by many staff officers, particularly those who had been involved in the 3rd Armoured Division trials. For this reason, as has already been noted, an initial purchase of some 25 off-the-shelf, non-TEMPEST approved laptop systems was placed with

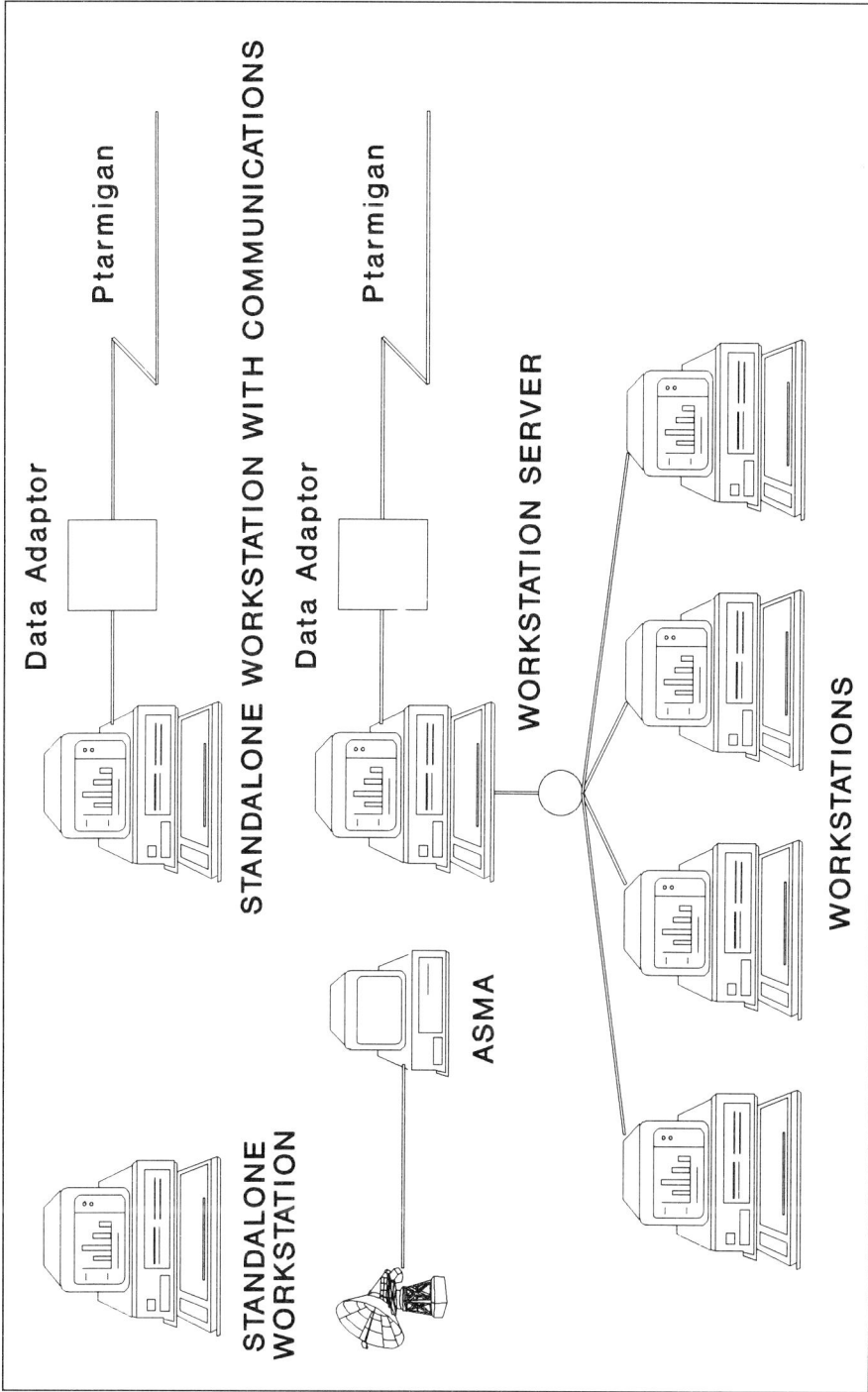

Fig 4.3 DICS Hardware Components

GRiD to provide a DICS training set in BAOR. Six instructors from the Defence ADP Training Centre in Blandford were sent over to Germany to provide familiarisation training on these equipments for staff who were about to be deployed to the Gulf.

However, the operational equipments had to be procured via a different set of procedures and other staff branches now became involved. This Urgent Operational Requirement (UOR) fell into the province of the Director of Military Communications Projects (DMCP) in the MOD Procurement Executive (PE) – now renamed the Programme Director Military Communications Systems (PD/MCS). It was government policy that competitive tendering had to be used wherever possible in defence procurements and to apply it properly MOD staff believed that specification of the term 'laptop' would not permit sufficiently fair competition. Clearly, for the same reasons, the UOR could not possibly specify GRiDs. As a result, the requirement specification for the hardware became changed from 'laptop' to 'portable' as the UOR was staffed through the MOD system and it was 'portable' that was finally specified to the PE.

DMCP felt that, under the circumstances, PE responded very well to this UOR. The specification was for four types of machine: laptops (now to be read as portables) and desktops, both ruggedised and non-ruggedised. An Invitation to Tender (ITT) was issued to five companies on the 12 October with responses returned by the 18th. The Tender Board sat on the 22 October and the contract was actually let on the 26th. By anybody's standards this was a rapid and efficient piece of work. As it happened, the eventually selected system was also found to be the cheapest. None of the equipments was made in the UK, leading to the thought that perhaps our own defence industry had now slimmed down too far, to the point where there was little or nothing immediately available on the shelf.

The EUCEs and DEUCEs

Despite strong representations from the staff in 1st British Corps for the laptop computers, which included the sending of a staff officer over to MOD to plead their case, the competitive tender was placed with the word 'portable' as the requirement. The successful supplier turned out to be a company called C3 Secure Systems from the United States. The GRiD proposal had fallen because it had no expansion board capability and it could not be connected to a fibre optic local area network (LAN) without an adaptor box, hardware features that were required in the UOR. C3 Secure Systems offered two candidate computers: the End User Computing Equipment or EUCE, which was intended as the desktop system, and the Downsized End User Computing Equipment or DEUCE, which was to be the ruggedised 'portable'.

These computers had some advantages. The DEUCE was fully ruggedised and both EUCE and DEUCE had been designed for use with the US Armed Forces. The EUCE table-top system was based on the Intel 80386 chip and the DEUCE 'portable' used the Intel 80486 chip. They were very powerful

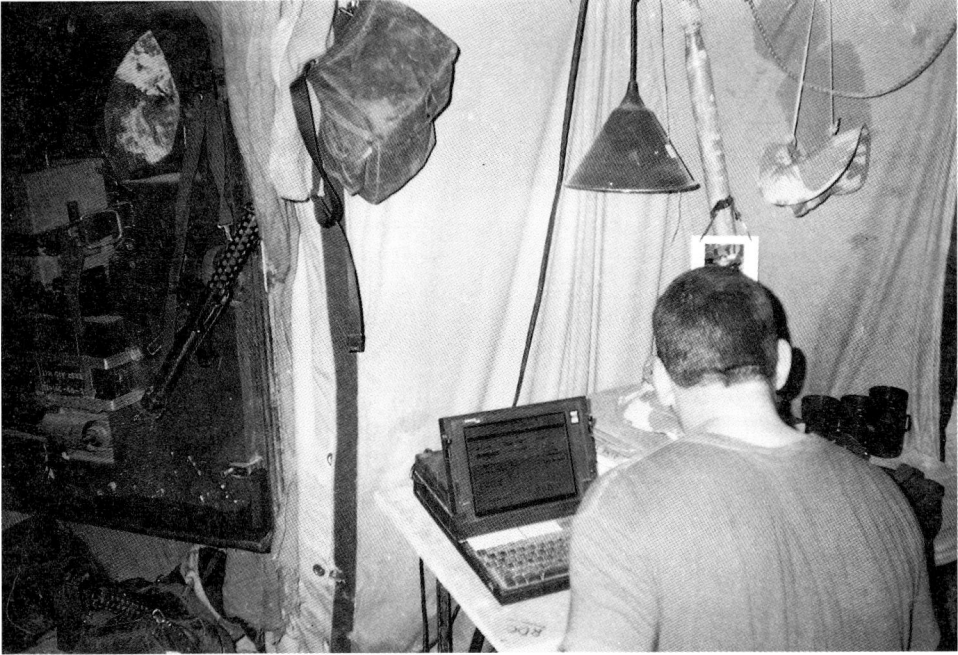

FIG 4.4 A GRiD Laptop Computer

systems and they made use of removable hard disks, a significant security feature. As an additional advantage, staff officers were able to carry with them, from one location to another on change of control, their complete data sets on the removable hard disks.

The DEUCE, as the 'portable', weighed some 40 pounds; the EUCE required either 115 volts or 28 volts power supply as did the printer that came with it. An additional large, heavy power pack was provided to convert from the 240 volts normal supply provided in UK field headquarters. For the vehicle installation kit, a further 10-pound invertor was required which made a particularly intrusive buzzing noise when in use.

These difficulties had all been highlighted in the DICS User Trials, the conduct and findings of which we describe in a later section.

Because of their ruggedisation, TEMPEST hardening and, to a lesser extent, their high capability (considered highly desirable to allow networking), EUCE and DEUCE were both relatively large for portable systems. EUCE was cumbersome to move when relocating the mobile command post (CP), and, as a portable, DEUCE was not the sort of flexible briefcase-sized laptop which could be tucked under the arm and moved from one staff cell to another. For some, this lack of flexibility was not fully compensated for by the systems' other features and, after 76 EUCEs and 41 DEUCEs had been bought, the case was made for the procurement of a number of true laptops. At the same time, the US Defense Act had been invoked by the US Government and the supplier

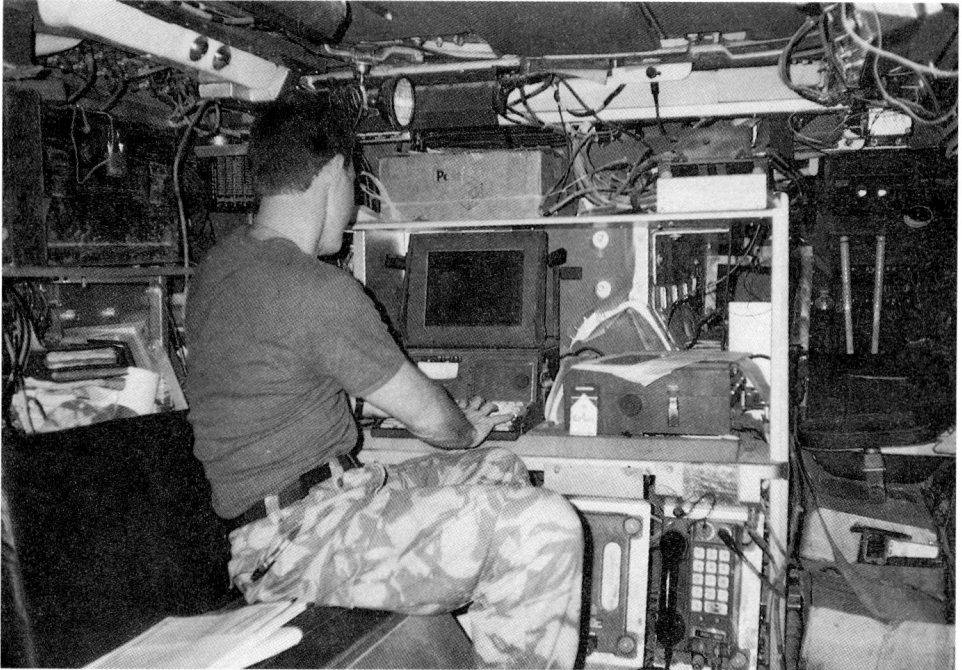

FIG 4.5 The DEUCE Computer

of EUCE and DEUCE, C3 Secure Systems,was under a *prima facie* obligation
to give priority to meeting the requirements of US forces.

The Second Purchase of Computers

For all these reasons, the second purchase of computers was based on
GRiD laptops: 102 TEMPEST-approved ruggedised systems, this number
being limited by what was available, and 39 non-TEMPEST-approved
versions. A further 106 DEUCEs were procured at the same time. As these
numbers could not be provided in the required timeframe and were in any
case still insufficient to meet the intended scalings, a further 39 grey plastic,
non-ruggedised, non-TEMPEST approved, commercial version Toshiba laptops were eventually purchased ahead of delivery of the GRiDs.

The total planned scaling of DICS computers for the brigade was 136; for
the entire division it was 434. By the start of the operation some 350 computers had been deployed, many of which were in the FMA and the FFMA
where they were used primarily for logistic purposes.

Design of the Software

Having established a basic hardware architecture for DICS and set in
motion the procurement process for the purchase of the systems, the HQ CIS

FIG 4.6 A Toshiba Laptop Computer

Group as the DICS Development Team now concentrated its efforts on the software design and its implementation. For all the good reasons outlined earlier, they had decided to base the software on the MSDOS operating system and to use standard packages that were familiar to the staff, wherever possible. The major software packages that were utilised are those shown at Figure 4.7

There was not a lot of time for a thorough analytical study. The system was intended to support operations initially at brigade, and subsequently at divisional level, and to cover in part those areas of battlefield CIS that are assigned to the *Wavell*, SIMS, LOGCIS and BICS/UNICOM projects.

It was not intended that DICS should cover operational command and control at unit level below the battlegroup headquarters. The design was based on the understanding that the information requirements of the force were as defined by Army Formation Standing Operational Procedures (AFSOPs) and the standard Army formats for Operational and Administrative Orders as modified for Operation GRANBY. In addition, the team interacted directly with the staff users in Germany to find out to what uses both the official and the individual personal computers were currently being put in the Headquarters. Team members discussed with staff officers the applications that they thought they would need in the Gulf, particularly those that could be provided by means of simple pre-prepared word-processing documents, spreadsheets and databases.

- **MSDOS** - Operating System

- **UNIPLEX** - Office Automation

- **SUPERCALC 5** - Spreadsheet

- **PCFILE** - Database

- **INFORMIX** - Database

- **INSTAPLAN** - Office Management

- **AUTOSKETCH** - Graphics

- **LANSMART** - Local Area Communications

- **KERMIT** - Wide Area Communications

FIG 4.7 The Major Software Packages

The priorities that the software development team worked to were first to implement a series of pre-formatted word-processing documents, then to make up brief formatted spreadsheets and, finally, to design simple databases; the last being used only where it was thought that the word-processor or the spreadsheet would be inadequate for the task. The pre-formatted word-processing documents included support for the standard military report writing procedures, JSP 101, which were adapted from those in use on CORMIS; standard formats for orders taken from the 'Staff Officer's Handbook', and AFSOPs (Part 2), Reports and Returns. Spreadsheets were designed to assist individual staff officers in managing resources and the database applications were used for more complex resource-management tasks.

Some problems were noted in practice with the software packages. Users who had experience with their own PCs tended to prefer more modern word-processing packages, such as WordPerfect 5, to that of the official UNIPLEX standard though its use was continued throughout DICS for all word-processing applications. UNIPLEX was also considered unsuitable for the spreadsheet applications and SUPERCALC 5 was brought in for this purpose. Although there is an integrated card-index file sub-system in UNIPLEX, this was not found to be adequate for most database purposes. The INFORMIX database system was thus included for this reason. However, INFORMIX was found to be too complex to use for the majority of applications, requiring a complete Structured Systems Analysis and Design Methodology (SSADM) study to obtain full benefit from it. In order to provide a simple but useful database facility, PCFILE was then added to the package repertoire by popular demand.

Two weeks after the first UOR had been signed on 20 September, Version 1 of the DICS software was delivered. It consisted primarily of the Army Formation Standing Operational Procedures (AFSOPs) Reports and Returns as pre-formatted word-processing documents, together with some support for the standard military report writing procedures, JSP 101, the latter having been taken straight off CORMIS.

The Software Architecture

A modular approach was taken to the software design. The overall concept was that a standard Human Computer Interface (HCI) should be presented to the user at the top level of the structure by means of a menu system implemented using UNIPLEX.

This approach had the distinct advantage that it would already be familiar to existing users of CORMIS; though it should be remembered that by far the majority of staff within the Division and the two Brigades had no experience of CORMIS, which was based at Bielefeld. At the next level down in the architecture came the package applications in three main categories: non-operational staff duties, operational staff duties and operational applications. Below that was the application environment, the computer system management suite, the local area network module and the wide area network module (see Figure 4.8).

The non-operational category made available general word-processing facilities, using UNIPLEX, for the writing of memoranda, briefs, formal and

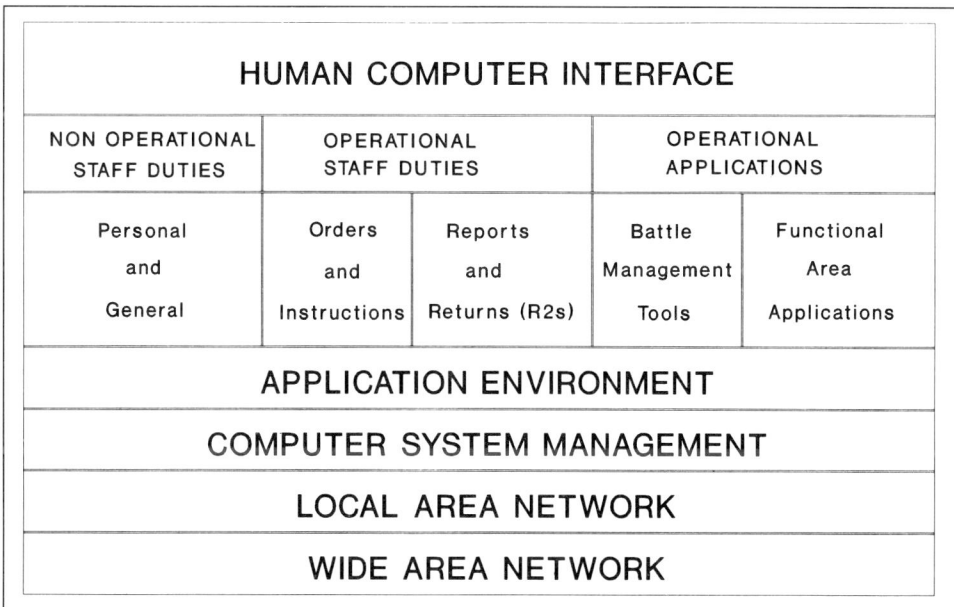

HUMAN COMPUTER INTERFACE				
NON OPERATIONAL STAFF DUTIES	OPERATIONAL STAFF DUTIES		OPERATIONAL APPLICATIONS	
Personal and General	Orders and Instructions	Reports and Returns (R2s)	Battle Management Tools	Functional Area Applications
APPLICATION ENVIRONMENT				
COMPUTER SYSTEM MANAGEMENT				
LOCAL AREA NETWORK				
WIDE AREA NETWORK				

FIG 4.8 The Software Architecture

routine letters, minutes and so forth. It gave some automated assistance to the user in that whenever the category was first entered, a file record card would be completed containing that user's full and abbreviated names and appointment addresses. Thereafter, every time the user typed in a memorandum, the system would automatically incorporate the correct abbreviated name and address and, for a letter, it would provide the correct full name and address.

The operational staff duties category initially included only an Operation Order and an Administrative Order under the Orders and Instructions subcategory, though this was subsequently expanded with other kinds of order, together with a wide range of Annexes. The different kinds of Report and Return, taken straight from AFSOPs, eventually numbered some 130, though few were much used in the Gulf simply because the staff had not had any opportunity to practise with them. This set of facilities was also implemented through formatted word-processing documents under UNIPLEX.

The Battle Management Tools and Functional Area Applications were largely predefined spreadsheets set up with SUPERCALC 5. Applications such as Combat Supplies Calculations, Brigade Movement Order Calculations and Staff Tables were among the 18 or so spreadsheets provided. In addition, a project-planning facility using INSTAPLAN and a limited graphics capability using AUTOSKETCH were provided on some systems. A summary of the key facilities, by function, is shown at Figure 4.9.

The Computer System Management module utilised UNIPLEX and the Local Area Network and Wide Area Network communications were based on LANSMART and KERMIT, respectively.

- OPS   - Operational Orders (OpOs)
            Fragmented Orders (FragOs)
            Reports and Returns

- MOV   - Formation and Unit Movement Plans

- ARTY   - Target Listing

- NBC   - Equipment Holdings

- MAINT   - Availability Return

- G1   - Personnel Reports
            Local Orders
            Nominal Rolls

FIG 4.9 Summary of Key Facilities

Regulated Anarchy

Not all the applications were developed direct by the DICS Software Development Team. The reader may recall earlier reference to the provision of a Field Records database on Mapper for the purposes of validating casualty details at the hospitals. Mention was there made of a similar DICS facility being developed as a backup to the Mapper capability. This DICS version was, in fact ,written in DBASE III by a padre, thankfully unemployed, who was whiling away his time in 33 Field Hospital at Al Jubayl. The source code was sent from the Gulf by modem back to the DICS Development Team who then, after thorough testing, incorporated it into the next software version for issue back to the Gulf.

This approach, which permitted the incorporation of talents and good ideas direct from the users, was referred to by the Head of the HQ CIS Group as 'regulated anarchy'. It was a policy that was intended to establish a necessary belief that the system really did belong to the users and not to the procurers or developers.

Other packages that were incorporated from outside the DICS Software Development Team included a knowledge-based system from Porton Down and an NBC program from Denmark. Then, just three days before the ground campaign started, concern was expressed in the Gulf about the problems of British fatalities who might have to be buried in Iraq. In just a 48-hour period, a database for the location of allied war graves was written by the Forward Support Team in the Gulf, was sent back to the UK for testing, quality assurance and configuration control, and then was re-issued forward to the units just as the Division started its advance.

ACISA and the DICS User Trials

Establishment of ACISA

During this period, as the DICS Development Team was coming into being and commencing its work, so a new organisation was being formally established as part of the long-term Army CIS strategy. This organisation, the Army CIS Agency (ACISA) based at Blandford, was formally approved by MOD in September 1990 and thus, as the DICS project was getting under way, so ACISA was beginning to establish itself in a new role with new staff in new accommodation.

Head ACISA, quite understandably, did not want to become distracted from these essential activities by any major involvement in DICS. He also felt that there were plenty of capable people already involved in DICS and 'the fewer fingers in the pie, the better'. On the other hand, it was clear to him that DICS could be establishing the way forward for Army CIS for years to come, and this was very much the new Agency's business.

The Role of ACISA in DICS

In the event, the Director of CIS (Army), DCIS(A), tasked ACISA with the establishment of a DICS System Support Team and the responsibility for oversight of configuration control, project documentation and the quality assurance of DICS. As part of this responsibility, ACISA would carry out the DICS User Trials.

The ex-CORMIS team leader who, as we indicated earlier, had been posted from Bielefeld to ACISA, was immediately employed by Head ACISA to run the DICS System Support Team and to act as the configuration control officer for DICS. The lieutenant colonel, who had been responsible for the testing of some of the 9 ID systems in the United States and who was now also based at ACISA, was given the task of setting up and running the DICS User Trials.

The Trials Philosophy

The DICS User Trials were planned in five phases as summarised in Figure 4.10. Phase 1 was to be the testing of the DEUCE and GRiD laptops together with the main user functionality, and this was to take place at the School of Electrical and Mechanical Engineering at Arborfield. Phase 2 was to be the testing of the EUCE desktops and the remaining user functionality, also to be carried out at Arborfield.

Phase 3 was to be electromagnetic compatibility (EMC) and TEMPEST testing with the *Ptarmigan* communications system, and this was to be carried out at the School of Signals at Blandford. Phase 4 was to be vehicle installation testing of the DEUCE systems, partly carried out at the School of Infantry at Warminster and partly at the School of Signals at Blandford.

- **Phase 1 - Testing of DEUCE and GRiD Laptops**
  Main user functionality

- **Phase 2 - Testing of EUCE Desktops**
  Remaining user functionality

- **Phase 3 - EMC, TEMPEST and PTARMIGAN testing**

- **Phase 4 - Vehicle installation of DEUCE**

- **Phase 5 - LAN and WAN integration testing**

FIG 4.10 Summary of Trial Phases

Finally, Phase 5 was to be the local area network (LAN) and wide area network (WAN) integration testing, also to take place at the School of Signals at Blandford.

The aim of these user trials was 'to assess, as far as possible in the time available, whether DICS as delivered, together with user developed software meets the user requirement and, in particular, is sufficiently simple, secure and robust for use in the Gulf'.

Conduct of the Trials

To this end, a comprehensive DICS Test Plan was drawn up, together with the test schedules, a failure policy and the evaluation criteria. Trials were carried out by teams of representative users under controlled conditions who exercised the test schedules in accordance with the test plan. A DICS User Trial Evaluation Panel was formed to evaluate results from the trials and to recommend actions. The findings of this panel were summarised in two main trials reports.

The Trials Findings

A number of important issues were raised by these trials. They demonstrated that the ruggedness, hardness and capacity of EUCE and DEUCE entailed a corresponding penalty of greater weight and bulk and, hence, of reduced mobility. The importance of being able to connect the computer systems together over the field communications circuits was strongly stressed and the development team was also urged to make improvements in the user friendliness of the DICS manuals, which had been produced at very short notice. The trials also showed a need for individual users to access the *Ptarmigan* wide area network (WAN) direct.

The DICS Management Structure

By this time the *ad hoc* development team arrangements, that had been brought together by the determination of the Head of the HQ CIS Group, had evolved into a working management structure that, together with parts of ACISA, eventually became formalised as the Desert Interim CIS Organisation. An outline diagram of this management structure is shown at Figure 4.11.

Altogether some 47 people were involved in the final organisation. Although the Project Coordinator (the original Head of the HQ CIS Group) supported the idea of a Project Assurance Team and a User Group, neither ever really got going properly. This was because many of the staff officers earmarked for these elements had to be 'double hatted' and were more urgently needed in their alternative roles.

The Project Office benefitted from the advice of a TA officer, who, in civilian life was the Training Manager for a computer consultancy firm and he

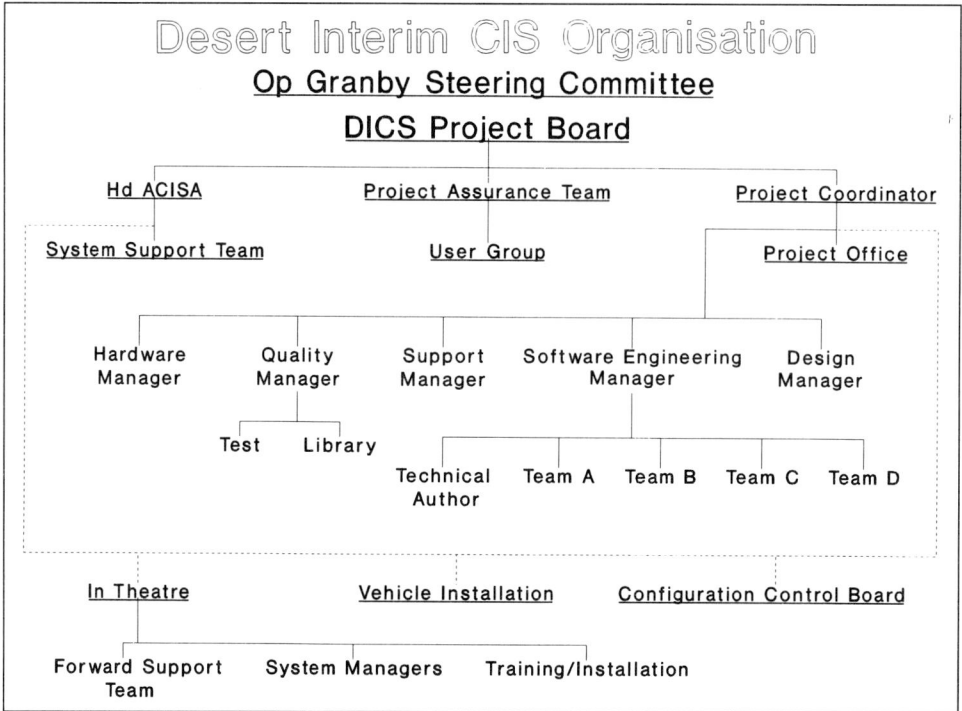

FIG 4.11 The Desert Interim CIS Organisation

spent much of his time writing the procedures manual. The Hardware Manager was based in the MOD and he tracked the procurement of the computer equipments. The Quality Manager was assigned from the Arms CIS Group, referred to earlier in the chapter, and he managed both a small Test Team and a Configuration Control and Library Team. The task of the Support Manager was to be in daily contact with the Gulf both by telephone and by data link so that files could be passed backwards and forwards between the support organisation and the users.

The Software Engineering Manager was the *de facto* designer of DICS and manager of the Main Support Team which was eventually established in a disused garage at Blandford. It was he who poured common sense on some of the wilder ideas suggested to the Team. He instituted the rules for documentation and testing and established a system based on an Informix Database to keep track of the multitude of concurrent tasks approved under priorities set by the daily Change Control Board meetings. He imposed a formal configuration control mechanism on the project. He also oversaw the work of the technical authors and of three small software development teams, one of which carried out the functions of systems integration.

The fourth team shown on the diagram (Team D) never became established. Change control requests received from the users were assessed every evening by the Change Control Committee who then tasked the teams for the following day according to available resources and the users' priorities. The evolution of the design of the complete DICS system was overseen by a Design Manager from ACISA. He was that ex-CORMIS team leader, whom we met earlier, who had such a wealth of practical software development experience, but who had managed to slip through the net cast by the Head of the HQ CIS Group. He was now, by courtesy of the Head of ACISA, able to bring his considerable experience and skills to bear on the development of the DICS project.

The main tasks of the System Support Team, run by ACISA, were those of configuration control and fault analysis, with some further software development work occasionally being required. A Configuration Control Board, also run under the auspices of ACISA, examined each new version of software and its documentation before it was issued to the Gulf to ensure that all procedures and testing had been carried out properly. The dates of issue for the four major versions of software are listed in Figure 4.12, though several intermediate versions were issued as well. The final version developed was 5.03 though detailed planning had been taken through to version 7.

In the Gulf itself, there were three key members of the DICS management organisation: an SO1 at HQ BFME in Riyadh, an SO2 in the FMA at Al Jubayl and an SO2 in 1 Division Main Headquarters. These three between them oversaw in-theatre the Training and Installation Team, the several Systems Managers and the Forward Support Team. The task of the Forward Support Team was to write and modify software actually in theatre and they had a data link back to the main programming teams in the United Kingdom to assist in this process.

- **Version 1** - **Delivered 2 weeks after first UOR was signed on 20th September 1990.**

- **Version 3** - **Developed for Brigade Applications; delivered 22nd October 1990.**

- **Version 4** - **Local Area Network and Ptarmigan use; delivered 15th December 1990.**

- **Version 5** - **Developed for Division Applications; delivered 15th January 1991.**

FIG 4.12 Major Versions of Software

The Information Management Strategy

Perhaps the most significant point to note in all of this is that no contingencies, had been made for any part of this organisation. The personnel were all either 'hijacked' from other CIS teams, or existing peacetime CIS agencies were prevailed upon to contribute personnel or to carry out specific taskings for the DICS Development and Support Teams. All the procedures and working practices had to be established from scratch; there were no SOPs to fall back on. A key factor in the successful outcome of this project was the enthusiasm and drive of the DICS 'champion', who perhaps did most to force the necessary organisation into being, though, it must be remembered, many others contributed as much.

It is thus very important that an information management strategy, defined by Michael J Earl[1] as the 'wherefore', be fully addressed when the information systems (the 'what') and the information technology (the 'how') strategies are being developed. Without the first, we lack the necessary means to put the other two into effect.

The Story Continues

We have now reached a stage in the story where the Division is training on the coast, some 300 km from its Assembly Area, the air war is about to start and preparations are in full swing for the ground war. We observe, in the next chapter, through the eyes of the SO2 at 1 Division Main Headquarters the introduction and use of the DICS system. How useful did it really turn out to be? What might have been done better?

---

1 *Management Strategies for Information Technology*, Michael J. Earl, Prentice-Hall, 1989.

5.

How We Got There:
II – Deployment and War

. . . air power can only do so much; the Army must go in on the ground to defeat the enemy's ground forces finally to win the battle.
Lt Gen (Retd) Edward M Flanagan, 1991

The 1st Division Deploys

The New CIS Staff Officer

In late December 1990, as the G3 CIS Staff Officer who had served with 7th Armoured Brigade and who had initiated so much of the early CIS work in the Gulf was on his way back to the United Kingdom to take up his studies on the Army Staff Course at Shrivenham, so his replacement was just finishing the Design of Information Systems Master of Science course at the same college. The new CIS Staff Officer would not be going to the Brigade, however. His appointment was that of SO2 G3 (CIS) in the Headquarters of the 1st Armoured Division, a post that had not been thought necessary before the emergency. As we shall see, it turned out to be essential and was one of the three new posts in the Gulf that had been assigned to the now official Desert Interim CIS Organisation which we described in the last chapter.

Into the Desert

On 5 January 1991, shortly after his arrival in the Gulf, the 1st Division moved out into the desert from its temporary base camp at Al Jubayl. The Force Maintenance Area, also located in Al Jubayl, and which had previously come under command of the Division, now came under HQ British Forces Middle East (BFME) in Riyadh. On 26 January, 1 (BR) Div was placed under the tactical command of the US VII Corps.

The Field Deployment of the Division

In common with its practice in Germany, the Division deployed three field Headquarters called Main, Alternate and Rear. Alternate was a fully equipped but only partially manned replica of Main and was used, as we have described in our previous book, as a Step-up headquarters. The step-up concept enables rapid redeployment of the tactical command and control functions of the field headquarters without causing loss of continuity, and provides for a measure of redundancy in the system facilities, thus reducing vulnerability. The Brigade HQs, however, did not use alternates in this operation. Although one of the brigades did have a vestigial form of step-up this was never employed as such.

The principal role of the Divisional Rear Headquarters was to control the logistic and support functions of the entire division. A separate Divisional Administrative Area (DAA) was also deployed. With the FMA at Al Jubayl, several hundred kilometres east of where the DAA was located, it was found necessary to deploy an additional support facility between the two in the general area of Al Qaysumah. This was known as the Forward Force Maintenance Area (FFMA) and is shown in some reports as 'Area Keyes'. A diagram of these deployments is shown in Figure 5.1.

To provide the Division with its trunk communications, there were *Ptarmigan* secondary access nodes at each of the Brigade Headquarters as

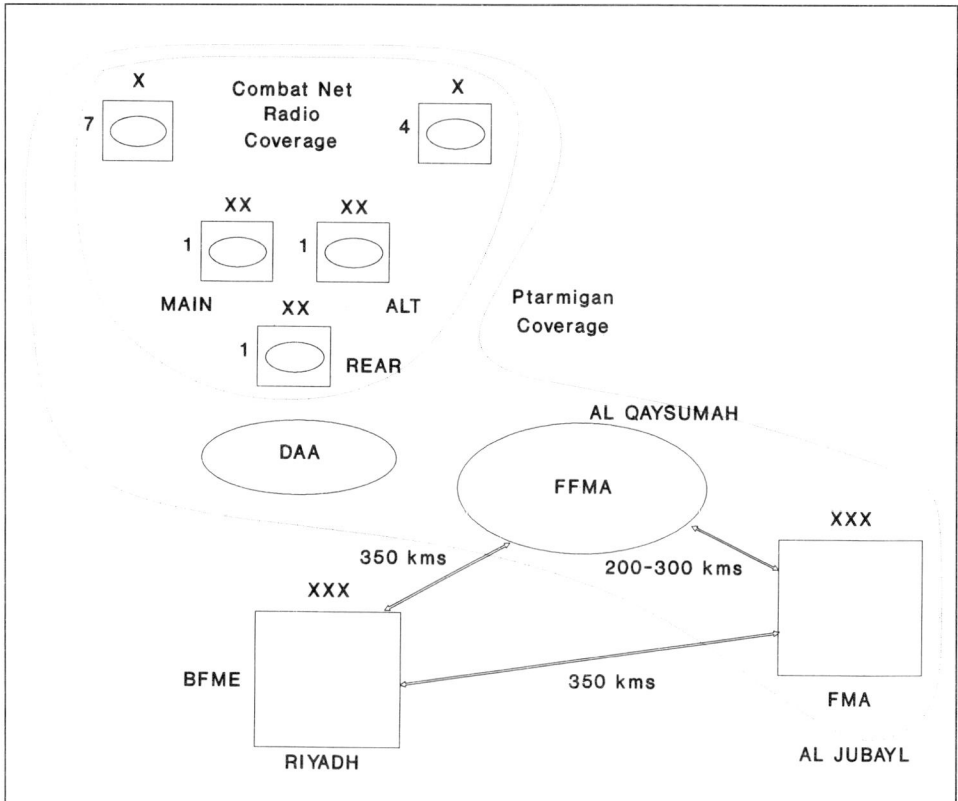

FIG 5.1 Field Deployment of the Division

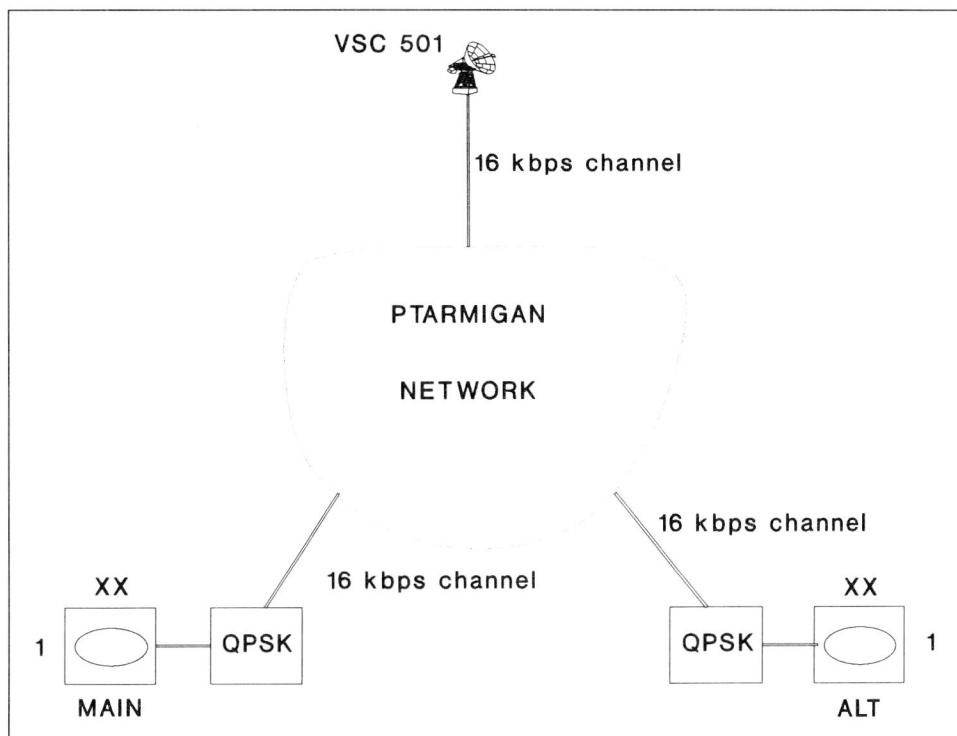

FIG 5.2 The Satellite Terminal at Main and Alternate

well as Main, Alternate and Rear Divisional Headquarters, and there was an additional node in the DAA. It was planned that VSC 501 satellite terminals would be placed at Main and Alternate to provide the strategic links. In the event, the satellite terminal from Alternate was repositioned at Rear and the original Main VSC 501 had to move on change of command from Main to Main. With the arrival of the Armoured CIS vehicles, the vulnerable VSC 501 detachment could remain in the DAA with the satellite communications facility being switched electronically between Main and Alternate Head-quarters through the *Ptarmigan* system as shown at Figure 5.2. In addition to these communications channels, there were the usual command nets using Combat Net Radio (CNR); High Frequency (HF) backup circuits and a number of special to role facilities, such as Ground to Air communications.

Tactical Layout of Divisional HQ

Diamond 1

The main command functions of the Divisional Headquarters were exercised from an area known as 'Diamond 1', which again followed the pattern practised in Germany. In the desert, Diamond 1 consisted of a group of Armoured Command Vehicles, type AFV 436, backed up to one another in a depression dug out by the Royal Engineers and surrounded by an earth or sand mound to provide some protection against shrapnel; these were known locally as berms and bunds. Draped tentage provided a single entrance to the complex and the whole area was covered by camouflage nets.

Even from a short distance away, it was quite difficult to detect the presence of the headquarters, as the photograph at Figure 5.3 shows. A number of 6 foot standard tables were placed between adjacent vehicles and in the centre of the complex the main planning and briefing area, known as the 'Bird Table', with its detailed maps was located. A diagram of Diamond 1 is shown at Figure 5.4.

Immediately to the left, on entering the complex, was the Communications Plans vehicle and next to that the Receipt and Despatch Centre (RDC) which housed the clerks. The photograph at Figure 5.5 shows these two vehicles with the 6 foot table in between and the RDC vehicle on the right. On the table is a DEUCE computer and alongside that is a photocopier, on top of which are the removable hard disks for the DEUCE, together with a laser printer shown just to its left. The RDC clerks did much of their typing at this table and the *Ptarmigan* electronic mail (Email) system for DICS and the one-to-one Email system to the US VII Corps bulletin board were also located here. The Communications Operations (Comms Ops) vehicle was not sited in Diamond 1 but was located with the *Ptarmigan* secondary access switch vehicles in a second quite separate complex known as Diamond 4.

FIG 5.3 Photograph of Diamond 1 Deployed Tactically

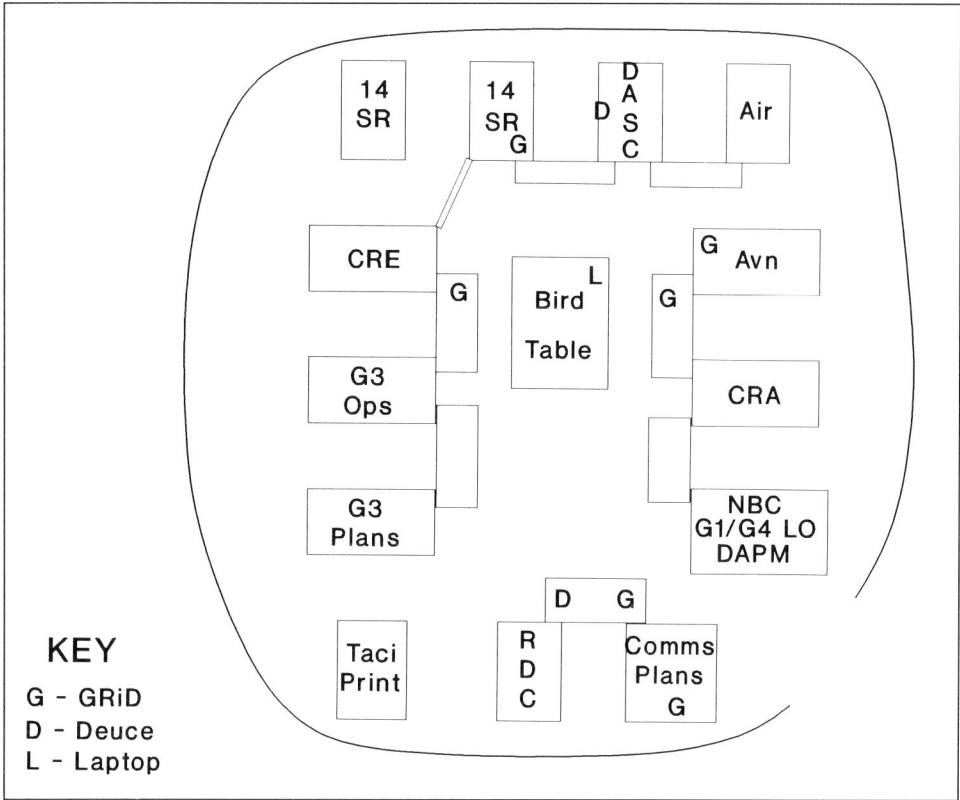

FIG 5.4 Plan View of Diamond 1

Beyond the Taciprint vehicle, which housed a high-resolution map and overlay colour printing facility, was G3 Plans, the Commander's vehicle which he shared with the Chief of Staff. G3 Operations (G3 Ops) was where the watchkeepers sometimes worked, but mostly they were to be found at the Bird Table itself. The table between G3 Plans and G3 Ops was used to plan future activities. The Commander Royal Engineers (CRE) vehicle housed the engineer staff and the table alongside their vehicle was used by them for their technical engineering work. There was a CRE communications net operating from the vehicle but this was not often used. Beyond that were vehicles from 14 Signal Regiment. This was a restricted area, curtained off by further tentage, which provided signals intelligence to the Headquarters. Alongside this was the Divisional All Sources Cell (DASC), which performed the G2 Intelligence function for the Headquarters and at the table outside this cell could be found Artillery Intelligence (Arty Int).

For most of the time, and throughout the war, an Air Liaison vehicle was positioned alongside the DASC. Next to that was the Aviation vehicle which, when the Air Liaison vehicle was absent, housed the Air Liaison Officer as

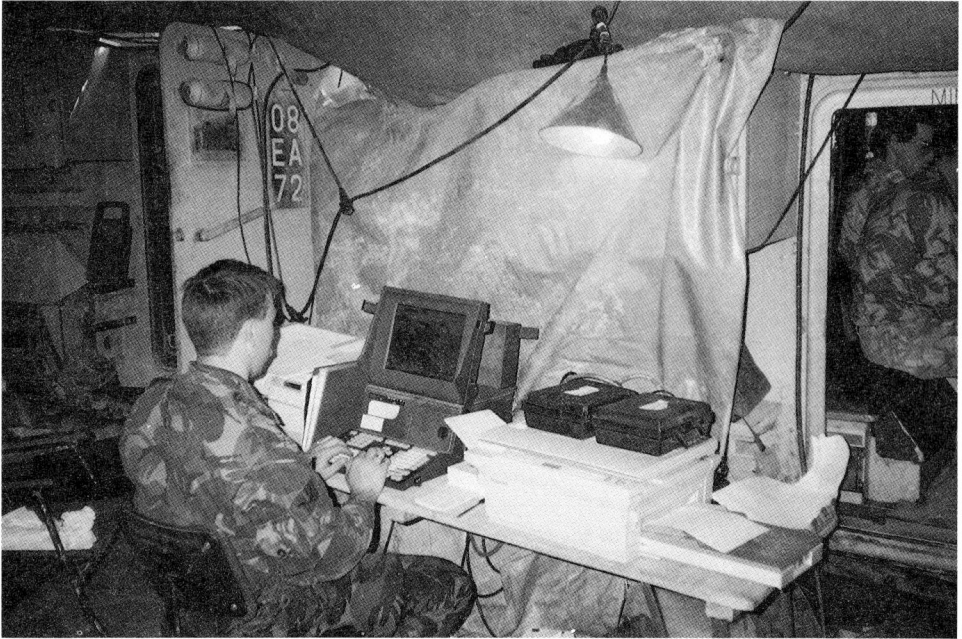

FIG 5.5 Comms Plans and RDC

well. The Commander Royal Artillery's (CRA) vehicle housed the gunner staff officers with an Artillery desk at the table outside. CRA had a gunner communications net in his vehicle which was 'remoted' to the external table. The final vehicle in the complex was used for a number of different staff functions: Nuclear, Biological and Chemical (NBC); G1 Personnel and G4 Logistics Liaison Officer (G1/G4 LO) and the Provost Marshal's staff (DAPM). The table between that cell and the CRA was used by the SO3 G3 Plans to plan all movements and effectively became the G3 Movements cell. The photograph of the table at Figure 5.6 shows the CRA vehicle on the left with the NBC–G1/G4–DAPM vehicle on the right. On the desk are two GRiD computers: the left hand one is a 1537 which is the TEMPEST-approved variety and on the right is a non-TEMPEST-approved 1530.

The Bird Table was located in the centre of the headquarters complex and provided the focus for all current operations matters. The Commander was briefed formally at the Bird Table twice each day, using the current operations map that was continually updated with felt-tip pens, and all operational decisions throughout the operation were taken at the Bird Table so that all members of the staff were aware of the factors that impinged on the decision. The Bird Table was manned by the SO2 G3 Ops and two watchkeepers who could monitor the divisional command net and answer the *Ptarmigan* telephones. An intercom was provided for the watch-keepers to update the staff in the vehicles throughout the Headquarters.

FIG 5.6 G3 Movements

Diamond 4

The supporting complex of 'Diamond 4', some hundred or so metres away, was very much simpler. By the start of the ground war, it consisted of five armoured vehicles which were dug into the desert with a surrounding earth mound and camouflaged in a manner similar to that for Diamond 1. A typical layout of these vehicles is shown in Figure 5.7.

The activities of the Diamond 4 complex had, for the most part, in the past been exclusively concerned with Royal Signals communications matters. The Royal Signals squadron commander responsible for communications into and within the Divisional HQ had his Communications Operations (Comms Ops) and Administration Operations (Admin Ops) vehicles located here. It was from these that he controlled the communications and logistic support for the Headquarters including the operation of the *Ptarmigan* Secondary Access Switch Message Centres (SAS(MC)). He also planned and executed the Divisional Combat Net Radio Rebroadcast Station plot for the four divisional VHF Radio Nets.

With the introduction of a new CIS facility, in the form of DICS, into the Divisional Headquarters, and the arrival of a newly appointed SO2 G3 (CIS), the Chief of Staff initially decided that the staffing of what at first sight appeared to be a very technical function should also be located in Diamond 4. At the outset, there was no vehicle available for the CIS staff officer and he therefore had to work from a small lean-to tent in Diamond 4 which

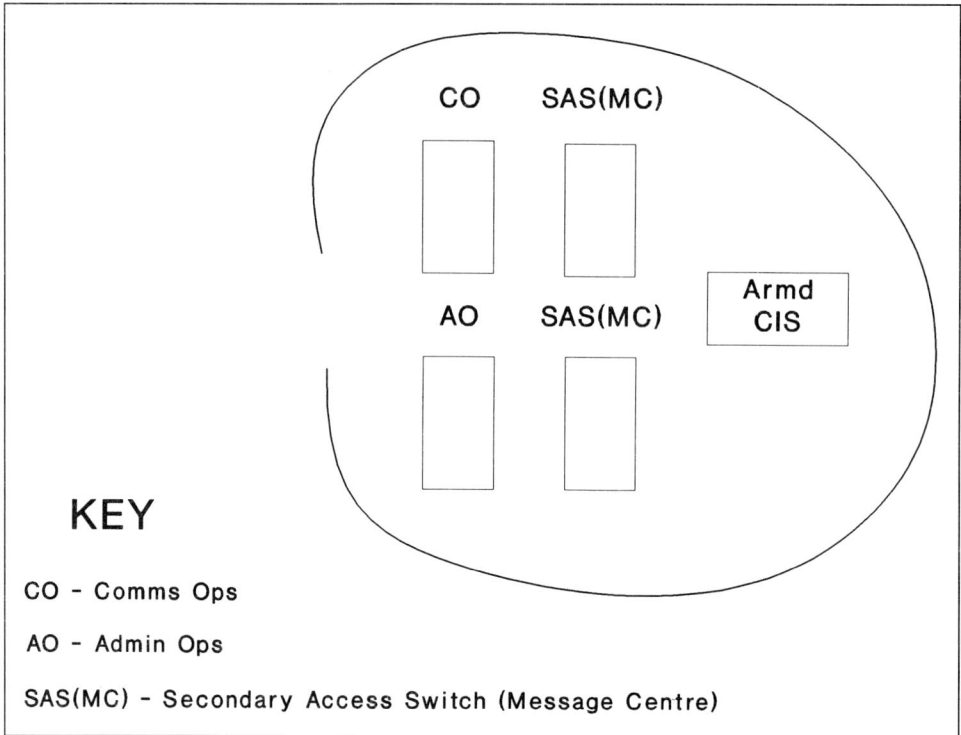

```
                    CO      SAS(MC)

                    ┌──┐    ┌──┐
                    │  │    │  │
                    │  │    │  │
                    │  │    │  │         ┌──────┐
                    └──┘    └──┘         │ Armd │
                    AO      SAS(MC)      │ CIS  │
                                         └──────┘
                    ┌──┐    ┌──┐
                    │  │    │  │
                    │  │    │  │
                    └──┘    └──┘

        KEY

  CO - Comms Ops

  AO - Admin Ops

  SAS(MC) - Secondary Access Switch (Message Centre)
```

FIG 5.7 Plan View of Diamond 4

was provided for and therefore shared with the Signal Squadron duty technician.

This arrangement was not ideal. In addition to being on call day and night, a situation that arose because he had no step-up counterpart to deal with the many CIS problems that occurred at all hours, the CIS staff officer was actually required to be assisting the Main HQ staff in Diamond 1, some hundred or so metres away in a separate complex. Having to move frequently between Diamond 1 and Diamond 4, particularly at night in a fully operational setting, was neither easy nor particularly advisable. There were also inevitable environmental constraints on arrangements for the support of DICS. Frontline support for the DICS software was supplied from the technicians' tent in Diamond 4, though there were more comprehensive facilities available at the FMA, while more substantial difficulties could be referred back to Blandford. In an ideal world, comprehensive facilities would have been provided in-theatre, close to the user, complete with the capability for testing and analysis; but planning had allowed neither for the introduction of a completely new CIS system at such short notice nor for its large-scale application in a theatre far removed from established support facilities.

The Armoured CIS Vehicle

At the start of the deployment, the Air Staff Management Aid (ASMA) for the Divisional Headquarters operated over a satellite channel that was provided by the VSC 501, which we referred to earlier, that moved from Main to Main on change of command. The vulnerability of this Land Rover-borne system was considered too great for a link of such strategic importance and so a case was made for an Armoured CIS vehicle to be deployed in Diamond 4.

It was agreed that two armoured 430 series vehicles, one for Divisional Main and one for Divisional Alternate, would be commissioned and that they would each contain a complete ASMA station. This would then permit ASMA to be provided in the way already shown in Figure 5.2. It was also agreed that the vehicles could be used to provide the much needed support facilities for the SO2 G3 (CIS) staff officer. They would be known as the 'Armoured CIS' and are shown as the fifth vehicle on the Diamond 4 diagram (see Figure 5.7)

The original plan was that the ASMA equipment would be installed at one end of the vehicle, allowing the space down the side to be used by the CIS staff. Unfortunately, this concept was not passed on to the installation team in the UK, who used most of the available space in the centre of the vehicle for ASMA, apparently not appreciating its dual role. A photograph of the inside of this vehicle is shown at Figure 5.8 and, once again, the CIS staff

FIG 5.8 The Armoured CIS Vehicle

found that this was not a particularly effective working environment for their role. The photograph shows a Wyse terminal which was used for setting up the ASMA link, the ASMA controller terminal and, on the right hand side, a rack of multiplexers. The empty slot between the two metal shields was intended to hold an ASMA printer, but as this was rarely used the CIS staff had it removed and positioned their laptop computer and printer in its place.

In fact, the vehicles were not delivered until just before the ground war started and, by that time, the CIS staff officer had through necessity become firmly based in Diamond 1, working from the table between the RDC and Comms Plans vehicles shown in the photograph at Figure 5.5. This turned out to be much more effective than the Diamond 4 location. He had direct access to the operational staff and, rather more importantly, they had direct access to him, so whenever they encountered difficulties with the CIS systems he was immediately on hand to help. As was mentioned earlier, the electronic mail link to the US VII Corps was sited at this table, using a 1550 GRiD computer fitted with a Hayes-compatible modem. This system gave access to the US VII Corps Email and bulletin board through Procomm software provide by the Americans. The availability of this bulletin board depended on the link to the American trunk system either by satellite or by radio relay. It could not be passed through *Ptarmigan* as the NATO interface is not transparent to data. Also positioned here was the DEUCE computer that was connected to the *Ptarmigan* data adaptor for access to the *Ptarmigan*-based Email system. The CIS staff officer was thus well placed to assist any staff who wished to send data in electronic form outside the headquarters.

Almost by default, this table became the G3 (CIS) cell. After the ground war was all over and during the post-operation debriefs, it was accepted that in future operations serious consideration should be given to locating a G3 (CIS) function permanently in the Diamond 1 complex. A suitable location might be to position the Armoured CIS in place of the Taciprint vehicle, which could well have been sited outside Diamond 1.

Main, Alternate and Rear

The Diamond 1 and the Diamond 4 complex were established at each Main and Alternate Divisional Headquarters location and re-established for every move of the Headquarters. Only one Rear Divisional Headquarters was set up for each move and this was slightly smaller in size than Diamond 1 and consisted in the main of the logistic staff cells.

DICS in the Headquarters

It was a deliberate policy that the scaling of DICS systems for divisional headquarters should be such that no computers had to be carried from Main to Main by the staff at change of control; all systems should remain with their associated staff vehicles, with only key staff moving from Main to Main. One of the attractive features of the C3 Secure Systems EUCE (End User Computing Equipment) and DEUCE (Downsized End User Computing Equipment) range was the provision of removable hard disks that would permit staff to carry classified data held on these disks from one headquarters location to the next without having to take the computers with them. However, as we described in the previous chapter, serious problems were encountered with the supply of the C3 Secure Systems equipments. It was eventually found necessary to supplement the EUCE and DEUCE computers initially with ruggedised GRiD systems and then subsequently with straight commercial laptops from Toshiba.

The Hardware Facilities in Diamond 1

The location of the DICS computers in Diamond 1 are shown in the plan diagram Figure 5.4. The C3 Secure Systems 386 EUCE and 486 DEUCE computers were all so-called 'near military specification' (near Mil Spec) TEMPEST-approved equipments (though not all agree that this was really true of the EUCE). There were three different versions of the 386 GRiD, also all near Mil Spec: the 1530 which was not TEMPEST-approved and had a fixed hard disk; the 1537 which was TEMPEST-approved and had a removable hard disk; and the 1550 which again was not TEMPEST-approved but had a Hayes-compatible modem fitted in the back. The Toshiba 3100s were all civilian standard non-TEMPEST-approved 386 laptops with fixed hard disks.

There was a GRiD 1530 computer in the Comms Plans vehicle together with a GRiD 1550 and a DEUCE on the table outside. These latter two have already been mentioned providing the US VII Corps with regard to Email and the *Ptarmigan* Email facilities. Alongside these was a Hewlett Packard *Laserjet* III laser printer and this was connected via a T-switch to both the GRiD and the DEUCE computers. All local printing for the headquarters was carried out on this printer.

As it happened, conventional wisdom had warned that the delicate mechanism of a laser printer would not survive the rough handling and dirty conditions of desert field use and so no such printers had been officially procured for the headquarters. However, this particular laser printer had been 'found' by the CIS staff officer on the quayside at Al Jubayl and he had commandeered it to meet the priority needs of the Division. It was used constantly by the staff of divisional headquarters throughout the operation, being carefully transported from Main to Main by the Chief Clerk, who was at pains to ensure its survival. As well as being a very useful aid to the staff, it was for him indispensable as a major saver of time. In the event, this laser printer continued to serve faithfully throughout the war and only at the very end did a line start to appear down all the printouts. The immediate thought by the CIS staff was a scratched drum, but on changing the toner cartridge the output was as good as new. This printer survived better than any other in the headquarters and provides one of the principal computer equipment success stories of the operation.

G3 Ops used a Toshiba laptop computer at the Bird Table and there were GRiDs in use by the Engineers, on the table outside the CRE vehicle, and by the Signals, inside the 14 Signal Regiment vehicle. This 14 Signal Regiment GRiD was damaged very early on while it was being powered direct from the vehicle power supply. Because of the shortage of GRiDs it was not replaced for some time, but the signals staff seemed to cope well without it. DASC had the second of only two DEUCEs that were used in the headquarters and this was mounted inside the vehicle where the *Wavell* terminal had previously been located. Aviation and CRA also had GRiDs as shown on the diagram. DAPM was issued with a GRiD but found little use for it and it was subsequently reissued as the one that was used by Comms Plans. The NBC cell had a Toshiba laptop that was not part of the DICS system. It provided a new NBC reporting facility which had been bought from Denmark, called BRUHN.

The Local Area Network

One of the key features in the Urgent Operational Requirement (UOR) originally specified for DICS, was that the computers procured should be capable of interconnection via a fibre optic Local Area Network (LAN). This was considered essential in order to permit local staff cells within a headquarters to share common data across a network, including for example, the facility for electronic mailing. It was the availability of a fibre-optic interface that helped to rule in favour of the C3 Secure Systems EUCE and DEUCE equipments and against the GRiDs when it came to the awarding of the initial contract.

In preparation for the introduction of the LAN, the CIS staff officer investigated what optical-fibre cable runs would be needed between the vehicles within the headquarters. He immediately encountered problems. There were no access points on the sides of the vehicles to which fibre-optic cables could be connected. Cable runs would either have to go through the back door of the vehicle or through the NBC over-pressure valve to connect to the computer inside. In both cases the cable run would compromise the NBC protection system for the vehicle when it was operating in its 'closed down' mode. With the spectre of chemical warfare a real possibility, the staff were naturally enough not particularly happy about either of these options.

In discussions with the divisional staff, however, the CIS staff officer found little support for the concept of a LAN. Even those who were keen and enthusiastic computer users felt that it was yet another level of unnecessary complication on a system that they were already having to learn 'on the job'. When they wanted to move data around, which would normally only be for printing, they could use standard 3½-in floppy disks. They saw little advantage in trying to network the system within the headquarters.

On the other hand, the Chief of Staff was a powerful advocate of the LAN concept. He felt that networking could bring real advantages to the work of the headquarters in two main areas. First, in the global visibility it would give to information within the headquarters so that all staff could be better informed, and secondly, in facilitating the drafting of operational orders by individual staffs directly adding their separate contributions to centrally held documents. There is little doubt that the Chief of Staff was correct in principle. The difficulty found in practice, however, was that the staff were not trained in using such methods and the team would need a lengthy work-up period to become proficient in applying the appropriate techniques. That time was not available to the headquarters staff in the field.

In the event, however, the whole issue became academic. As we have seen from the foregoing section, only two C3 Secure Systems computers were provided for Diamond 1: the DEUCE in the DASC and the DEUCE on the table outside RDC. Neither the GRiDs nor the Toshibas could be connected to a fibre-optic LAN. There seemed to be little operational point in connecting the DASC to the RDC by means of a box the size of a jerry can, especially since the All Sources cell made very light use of their DEUCE. The LAN facilities provided by DICS were thus not used at all in Diamond 1.

Although Rear Headquarters had more DEUCE systems than Main, the logistic staff tended to operate within their own functional areas and they too decided not to implement the LAN facility. A DICS Local Area Network was implemented at the FMA in Al Jubayl, but again, these were mainly logistic users and the local CIS staff reported little use of the network as such.

The Wide Area Network

Access to the Wide Area Network (WAN) was via the *Ptarmigan* data adaptor. This was not a network in the full sense of the term, but it did permit one user to dial up another user via the *Ptarmigan* trunk system and establish a circuit switched, one-to-one link between their computers. File transfer, message passing and Email were the principal facilities that were provided by the DICS software which made use of the KERMIT public domain package.

The original concept was that the Chief Clerk using the RDC terminal would control all wide area network access with users linking in through the LAN. Since the LAN was not implemented in the headquarters, this restriction was not imposed and any staff terminal with access to a *Ptarmigan* data adaptor could use the Wide Area Network facility. All the computer systems, but for the 1530 GRiDs, could interface direct to the *Ptarmigan* data adaptors which, in addition to the one installed in the RDC vehicle, were subsequently fitted to the CRA and CRE vehicles as well.

Security Issues

As a result of the equipment procurement problems and the consequent need to use non-TEMPEST-approved GRiDs and Toshibas within the

Division, a local waiver was granted by the G2 staff to the normal security rules concerning the processing of classified information within the headquarters. It was a considered judgement that the TEMPEST threat was minimal at Al Jubayl and virtually non-existent once the headquarters had deployed into the desert. The field headquarters was not an environment where electronic eavesdropping could take place without a very high risk of discovery and so authority to process data up to SECRET on non-TEMPEST-approved machines was granted. However, this was not the case for the passage of data over the communication systems where the normal security rules still continued to apply.

The Brigades and the Battle Groups

Within the two Brigade headquarters there were between three and six DICS computers, depending on the stage of the build up. These were often used, connected to the *Ptarmigan* communications network, for the passage of written orders between the Division, the Brigades and the battle groups.

Each battle group was initially issued with one GRiD and then, wherever possible, with a second. It was left to commanding officers and brigade commanders to decide where they should locate the systems within each of the battle groups. Many of the battle groups, on first receiving them, sent their DICS systems back to their peacetime locations in the Gulf where they were used to run administrative nominal rolls. Once an operational need was established, however, they were quickly returned to the field, where they tended to be used at battle group headquarters, connected into *Ptarmigan* via an SCRA link.

Staffing for DICS

In addition to the G3 CIS staff officer at Divisional Main Headquarters, there were four systems managers deployed within the 1st Division to look after the DICS systems. There was a captain based at Main, who also worked from the Armoured CIS vehicle in Diamond 4, and a warrant officer at Rear together with warrant officers at each of the two Brigade Headquarters. The system managers were responsible for providing direct support to all the staffs in their designated areas and for both hardware and software maintenance of the associated DICS systems. In the case of the Brigade systems managers, this included all those systems at battle group level and in the Brigade units, as well as at Brigade Headquarters, giving them each some 20 different locations to look after in all. The two divisional headquarters systems managers were responsible for the remainder of the divisional troops and had some 30 different locations to look after between them.

The systems managers were only able to carry out immediate-action first aid on the DICS systems in their areas of responsibility and provide some assistance to the staffs that they were supporting on how to use the systems. Typically, they could carry out basic fault finding on the computers and printers and advise on software packages, such as SUPERCALC spread-sheets. It was found in practice that their one-week course at Blandford was a quite inadequate preparation for the role that they were performing and without diagnostic software or hardware test equipment it is much to their credit that they managed as well as they did.

More comprehensive support came from HQ FMA at Al Jubayl, where there was a CIS staff cell set up with some 15 to 20 support staff for DICS. By the start of the air war this cell was being run by a second SO2 G3 CIS; that same staff officer who had some few weeks earlier been procuring the Field Records system as a member of the Corps CIS staff in Bielefeld. This CIS cell at Al Jubayl provided the focus for all the hardware, software and educational support effort for the DICS systems in the Gulf, a focus for the detailed work that was being carried out by the Desert Interim CIS Organisation in the UK, that we referred to in the previous chapter.

All new issues of DICS equipment to units were made direct by the CIS team from Al Jubayl. Although such issues were supposed to be channelled through the Divisional CIS staff, their initial lack of transport and limited staffing levels precluded this more normal processing through the chain of command. As a bonus, however, instructors detached from the Defence ADP Training Centre at Blandford were made part of the Al Jubayl team that did the issuing and they were able to give a brief two-hour, on-site training session about the DICS facilities to new users at the time that they received their equipments.

Maintenance facilities for the DICS hardware were very limited. One REME sergeant, based at Al Jubayl, provided the sole general repair facility for all the systems. The dusty, dirty conditions prevailing in the desert combined with user inexperience led to a high incidence of equipment failure.

Problems in Practice

The C3 Secure Systems Equipments

The findings of the first user trial had highlighted certain constraints on the flexibility of EUCE and DEUCE as a result of their relative bulk. The order, however, had by then been placed for the first purchase. Fortuitously, it became possible to add GRiDs to the order when DICS was extended to cater for the Division. The Toshiba laptops were at that stage also included, primarily for use by the systems managers.

Although the DEUCE uses an Intel 80486 chip and is, even by today's standards, a powerful machine, this was not evident to the user. In part this reflected the fact that it had not been possible to provide most of the users with the training required to derive the full benefit from the machine's capabilities; in addition, the UNIPLEX-based applications of word processor and spreadsheet did not run as quickly as had been expected and this is thought to be because UNIPLEX was designed to operate in a Unix environment. The overheads associated with converting it to run in an MSDOS environment may have contributed to the fact that the equipment did not perform to its full potential.

Problems with the Software

The DICS system managers both at Division and Brigade level found themselves with few tools and limited training, in continuous demand as the staff experienced a new set of problems with every new software update. The DICS equipment had initially been issued to the Gulf with its software at version 3.01. Within a period of four months, nine further updates had been issued to correct faults and enhance facilities; the final software version being 5.02.

There were many problems with the updates which were issued on floppy disks that automatically installed the new software version on to the hard disk of the computer system. Sometimes faulty installations resulted, particularly where the assumptions made by the software developers about the build state of the recipient computer systems were incorrect. The resulting software faults were often hard to track down.

When the first issue arrived, the rapidly produced software inevitably contained a number of faults and many intended and required user facilities not yet implemented. However, the DICS team in the UK were striving desperately to meet a set of difficult operational requirements within impossible timescales and this tended to preclude lengthy testing and evaluation. The result was that the staff in the field were seeing at times a limited service that was due to no fault of the systems managers.

Typical of the kinds of problem that the system managers had to cope with was the 'joined drives' issue. The DICS systems were built using the proprietary Microsoft Disk Operating System (MSDOS) version 3.3. This version of the operating system does not support fixed disk partitions that are larger than 32 Mbytes, a difficulty that has been overcome in later versions of the operating system from 4.0 onwards. The standard technique for working around this limitation when a hard disk larger than 32 Mbytes is used, is to define a number of logical drives all of which are less than this 32 Mbyte maximum. DICS was issued with a single physical 40 Mbyte hard disk drive that was partitioned into two logical 20 Mbyte hard disk drives identified as 'C:' and 'D:'. These were intended to be used for applications programs and applications data, respectively. However, in order to try to make the use of these two logical drives as transparent as possible to the user, the MSDOS 'join' command was employed in some applications. This unfortunately led to numerous software errors when the 'join' failed, often resulting in user files becoming 'lost' or distributed to the wrong drives and directories. Explanations about 'joined drives' and 'logical partitions' were not met with much enthusiasm or understanding by the average hard-pressed user.

With the start of the ground war imminent and these very real difficulties in mind, the CIS staff officer in the divisional headquarters decreed that Divisional Main and Alternate would not upgrade beyond version 4.01 of the software despite claims for the subsequent releases. Although this policy resulted in a number of different software versions being in use throughout the force (since complete systems were still being issued from Al Jubayl with the latest release of software current at the time of issue) it was nevertheless felt that the risk of a new software fault on going to war was too great for the headquarters staff to accept. Fortunately, all software releases were intended to be both upwards and downwards compatible, and in the event no compatibility problems between the different releases were actually noted.

Viruses

Another software problem that arose concerned the occurrence of computer viruses. Because DICS employed the widely used personal computer (PC) operating system MSDOS there was great potential for users to import their own locally purchased software on to the system. In many cases, this was software bought in Al Jubayl or Riyadh and this was considered to be at great risk from computer viruses.

A particular problem was *Joshi*, though other viruses were detected as well. *Joshi* is a virus that originates from India and is a so-called 'boot sector' virus in that it modifies the program code at Track 0, Head 0, Sector 1 of the infected disk, the code which is used by the computer as part of its initial start-up procedures. The observable effect of the virus is to demand that the phrase 'happy birthday *Joshi*' be typed in by the user before he was permitted to continue. However, another more serious side-effect was noted by the CIS staff officer. Although control had been returned to the user, after the phrase demanded had been correctly typed in, some parameters had also been changed internally, notably that of the floppy disk capacity. The computer now viewed the floppy disk drive as having half the capacity that was actually available. This meant that the computer could no longer read the full capacity floppy disks that were provided with the system. If the user carried out a formatting operation on a new floppy disk, this would be formatted at half its capacity; it could then only be read by a *Joshi*-infected system.

As a result of this, the CIS staff found themselves investigating a flood of complaints where users' floppy disks could no longer be read by their computer systems. Most of these cases were thought to be the side-effects of *Joshi*.

Virus detection software was available and in use. Sophos Sweep had been implemented on all the DICS systems but the earlier versions were not able to identify *Joshi*. Subsequently a range of different virus-detection and disinfection programs were incorporated into the software; these included Scan and Clean from McAfee, Norton Utilities, and the Sophos suite. All infected hard disks were thoroughly 'cleaned' and all infected floppy disks were physically destroyed, bringing the virus outbreak under control. It was clear, however, that more stringent measures would be needed in future if the risk of losing operational capability through the importation of virus-infected unofficial software was to be reduced to an acceptable level.

Printers and Disks

A range of six different printers was used within the Division. The CIS staff officer had noted that a printer was a truly essential part of the total system to the staff users; a computer without a printer was deemed to be no more than about 25 per cent effective.

Unfortunately, he also found that an unacceptably high proportion of the large green boxes that were the C3 Secure Systems 80-column printers, issued with the EUCE and DEUCE computers, failed with symptoms of what appeared to be burned-out motors. These were 110-volt systems which had to be run from special generators and they did not prove to be either very reliable or very useful. Whether the fault lay in the printer or in the power supply was never established.

The Diconix and the Datawatch printers, the latter being the TEMPEST version of the former, were both prone to paper-out errors and were felt to be not sufficiently robust for a field environment. The Genicom wide-carriage printer, which provided 150-column printout for logistics staff who needed that width for their large spreadsheets, was not found suitable for mobile PCs. The Canon *Bubble Jet* printers, on the other hand, were very popular and quite reliable. The only staff criticism about these printers was that they were issued without sheet feeders, which meant that they had to be stood over and hand-fed when producing multiple page documents. Appropriate sheet feeders were available from the manufacturer at the time but had not been bought.

The greatest surprise of all, however, was the Hewlett Packard *Laserjet* III printer that has already been mentioned. This gave excellent service throughout the operation; the only difficulty experienced being in the supply of its toner cartridges, which were eventually purchased locally from Riyadh. A physically smaller version, ruggedised and protected from dust would seem to form the basis for an excellent field-worthy printer.

In the main, the hard disk drives gave few problems but the floppy disks were a major source of difficulty. The industry standard 3½-inch double-sided disks, both normal and high density were used, to give 720 kbytes and 1.44 Mbytes capacity, respectively. Although this standard does have some protection in the form of a plastic case and a metal shutter, it was not sufficient to prevent dust from the desert getting on to the magnetic surface and permanently damaging many of the disks as well as the computer disk drive heads.

The least robust part of the DICS computer system was found to be the floppy disk drive. There was a clear need for many more disk cleaning kits to be issued and a standard cleaning process to be enforced, possibly by the computer software as part of a start-up user daily maintenance procedure. In addition, recognition that the staff are likely to carry their data disks around with them in their pockets suggests the need to provide small individual plastic protective cases for all disks.

Power Supplies

There was a significant increase in the demand for 240-volt AC mains power supply as a result of the introduction of the DICS equipment into the Division. The standard field generators used to provide this supply, however, often damaged certain items of computer equipment because of the all too frequent and severe power surges. The major casualties of this effect were the GRiD 1537 computers, the GRiD AC to DC transformers (known as 'silver bricks') and the Toshiba AC to DC transformers. The need, in addition, for 110-volt AC field supplies specifically for the C3 Secure Systems printer created further problems, notably the burning out of some such printers when inadvertently used on the 240-volt AC system.

The Practical Benefits of DICS

Primary Uses

Within the Division, the CIS staff officer perceived DICS as a sketchy command and control system for division, brigade and battle group staffs. It was used by the operations staffs primarily as a text word processor to write, edit and receive orders, and as a distribution system to disseminate those orders up and down the command chain by means of the electronic mail facility. Its principal use by the logistics staffs was for spreadsheet calculations, with files being sent backwards and forwards between the logistic cells of the field headquarters and the FMA at Al Jubayl. There were no DICS applications specific to the intelligence staffs, and apart from limited use as a word processor, they tended not to use their systems very much.

Word Processing

With no typewriters taken into the field, the staffs relied totally on DICS for the typing of all documents, even down to routine daily orders. Had DICS not been there, no doubt some other means would have been found, but word processing was nevertheless a very important feature of the system. The cohesive set of applications that permitted the con-struction of a large Operation Order (Op Order) within the system by individual staffs working together on separate parts was found to be of great practical value, even without the Local Area Network facility. The staffs became versatile at using this technique and, with more practice, would no doubt have developed it further into a slick and well established SOP. In an after-the-event assessment report of the Op GRANBY CIS facilities, it was considered 'unlikely that plans could have been prepared and disseminated in such detail and with such clarity without the DICS WP (word processing) facility. This had a direct impact on the success of the operation.'

Spreadsheets

The spreadsheet facility was also found to be most useful, though again not the UNIPLEX version. DICS provided two spreadsheet facilities: UNIPLEX and Supercalc 5. Few of the built-in spreadsheet applications that had been developed by the DICS team were fully used as they did not precisely meet the specific requirement. UNIPLEX spreadsheets were not used at all by those staff who really knew what they wanted. Most of the logisticians used Supercalc throughout, many of them building their own applications which often incorporated complex multiple spreadsheets. They were of immense value in many areas and an indication of their diverse use is given below, with a summary at Figure 5.9.

- Artillery Target List at 4th Armoured Brigade. Data were entered at Tactical Headquarters (TAC) and transmitted over DICS Email to the Fire Direction Centre (FDC). This cut out the need for long voice transmissions to pass the essential artillery target data;
- NBC Census of all units to determine holdings against scaling, which produced the theatre NBC equipment requirement and priority of issue by unit;
- Maintenance Availability Return prepared at Divisional Rear to show the availability of key items as specified by the Divisional Commander;
- Engine and Main Assembly Return prepared as for the Maintenance Availability Return;
- 7th Armoured Brigade Workshops staff tables which were used for operational moves in theatre and for resource management;
- Postal and Courier Accounting System.

- **Artillery Target List at 4th Armoured Brigade**

- **NBC Census**

- **Maintenance Availability Return**

- **Engine and Main Assembly Return**

- **7th Armoured Brigade Workshops Staff Tables**

- **Postal and Courier Accounting System**

FIG 5.9 Some Spreadsheet Applications

Most of the spreadsheet applications were stand-alone and thus could not be used for the automatic aggregation of information as it was passed up the logistic command chain. Even where spreadsheets were being used at a number of levels, results from one level had to be manually entered into the spreadsheet at the next level up. A real need for this kind of facility was identified in order to speed up the compilation of logistic returns as they are sent up the command chain.

Databases

Two special-to-task database applications were developed in the INFORMIX language by the DICS team, but these were little used. Some staff attempted to develop their own databases in the field using the much simpler 'card file' facility that is also provided with INFORMIX, but they found that this proved to be either too small or insufficiently powerful for many applications. Some of those applications that were successfully developed using the INFORMIX card file facility are shown at Figure 5.10.

As the full INFORMIX programming language facility was too complex for other than professional programmers to develop database applications, a user-oriented database called PC File was also provided with the DICS software. This was used to great effect in the field and some of the database applications developed by staff users with PC File are listed at Figure 5.11.

The advantages of an integrated word processor, spreadsheet and database environment, such as that provided by UNIPLEX, seemed in the Gulf to be outweighed by the more widely known, simpler to use and better liked features of the separate WordPerfect, Supercalc and PC File packages. Had users already been familiar and well practised with an integrated office facility, and had DICS provided a complete office system together with its communications, then the story might have been different. As it was, this was not the time for staff to have to learn new office management techniques.

- **Control of Officers Confidential Reports**

- **Part II Orders**

- **Unit Nominal Rolls**

- **28 Field Ambulance Monitoring System**

FIG 5.10 Informix 'Card File' Applications

- **Equipment Management for Royal Engineers**

- **Manifests for Recovery Operation**

- **Container Management for Supply**

- **Vehicle Call Forward and Statistics for 6 Armoured Workshops**

FIG 5.11 Some PC File Applications

Electronic Data Transfer

Electronic data transfer was used extensively throughout the operation and became almost indispensable. Although referred to as Email, the DICS software actually provided little more than a file-transfer facility. Nevertheless, this proved invaluable to the staffs. Both Op Orders and Fragmented Orders (FragOs) were sent through the DICS *Ptarmigan* Email as a matter of routine. The orders had some 15 addressees from Main Division, of which 12 were on the *Ptarmigan* network. By using *Ptarmigan* Email, FragOs were received in under two minutes by each addressee. All 12 had to be sent as separate transmissions because there was no broadcast facility built into the DICS KERMIT software, though such a feature was put high on the priority list of future requirements. The Email system to the US VII Corps utilised a shareware communications package called Procomm which was linked into a bulletin board in their headquarters. This provided much more of an Email facility than the DICS system. The CIS staff officer had only to send one copy of the 1st Division orders over the link to the VII Corps and then arrange for the bulletin board to transmit the details locally to all the relevant addressees.

To be on the safe side, all orders were also sent by courier, but the Email systems were very much quicker and proved to be robust and reliable whenever the circuits were available. Some difficulties were experienced when the *Ptarmigan* links became very stretched and tenuous as the Division moved first to the far west, then north to the border and then east again all the way into Kuwait. During the war itself, there were times when the 7th Armoured Brigade Headquarters in particular was no longer in contact with Divisional Main Headquarters over the *Ptarmigan* system, even when using SCRA. The CIS staff officer, however, noted few occasions when data could not be sent by *Ptarmigan* Email.

The Email application 'was the one facility that transformed DICS from a useful word processor and calculator, into a powerful and battle winning CCIS (Command and Control Information System)'. A comparison of the time taken to prepare and pass down the command chain short written orders such as FragOs and Warning Orders (WngOs), by the several means of delivery is shown at Figure 5.12. The values given include the time taken for the extraction of information and the drafting of lower level plans at each level of command.

The Chief of Staff of the 1st Division took the view that this dramatic reduction in time to disseminate orders and plans helped the Division to get inside the enemy's decision cycle; an effect, the reader will be aware, that gives reason for the term 'force multiplication' when used in the context of command and control systems.

Certainly, there were some very real benefits obtained from the Email system. With the initial thrust of the 1st Division halted, as planned, some way short of the Basra road, US VII Corps ordered the Division to cut this road at short notice. The task was given to the 7th Armoured Brigade who successfully achieved their objective. The Chief of Staff of the Division believes that had the facility for sending co-ordinating instructions to the Brigade over the DICS Email not been in place, it is quite possible that this objective would not have been achieved by the time the cease-fire at the end of the 100 hours took effect.

Benefits that are rather easier to prove and quantify come from some of the logistic uses. A return, crucial for the provision of equipment assemblies had to be sent daily from Divisional Rear to the FMA. This was initially sent by road, and as the vehicle was away for 36 hours, two were required to be in transit at any one time. The return was considered so critical that the road delivery was replaced by a daily helicopter run, a decision that resulted in a considerable increase in daily cost and the commitment of a valuable resource. However, once DICS Email became available, the return was sent electronically in minutes.

• **DICS Email**	**100 minutes**
• **Fax**	**155 minutes**
• **Telegraph**	**185 minutes**
• **Despatch Rider**	**325 minutes**

FIG 5.12 Comparison of Timings for Short Written Orders

In another case, an Intelligence Summary (Intsum) was prepared daily between the hours of 0800 and 1400 and then sent through the normal message traffic system from the FFMA to the FMA. It did not arrive, however, until 0800 hours the following morning, by which time it was out of date. Once again, when DICS Email became available, it was transmitted in minutes.

Graphics

Although the AUTOSKETCH drawing package was provided with the DICS software, it was not much liked and was found to be difficult and laborious to use, largely because of the lack of any kind of pointing device. Without a mouse, a lightpen or a rolling ball all the drawing operations had to be done with the cursor keys on the keyboard and this was a severe disadvantage. Nevertheless, great efforts were made to use the facility in order to prepare graphical task organisations and map overlays. Even though it was quicker to draw them by hand, they then could not have been transmitted out of the headquarters by DICS Email and would have had to have been sent by the slower and less satisfactory means of either fax or despatch rider (DR). A limited number of roller balls were subsequently issued, but much of the work done in the Divisional Main Headquarters on Graphical Task Organisations and map overlays was carried out by the System Manager using the cursor keys.

The Chief of Staff of the Division identified a specific graphics need to assist the Commander and his planning staff in the planning process. A facility was required whereby the Commander could develop plans and generate overlays by 'drawing' directly on the map, using a lightpen or similar device. The resulting dispositions and overlays would then automatically be entered into the computer for incorporation electronically into the OpOs or FragOs. Such a facility would be needed only in the Commander's vehicle and, possibly, at the Bird Table. A fully automated map, however, would not seem to be a real requirement.

Reports and Returns

Much effort had been expended by the DICS development team on the production of a reports and returns (R2) facility which was based on Army Formation Standing Operational Procedures (AFSOPs). As described in the previous chapter, there were some 130 different reports and returns that were eventually made available in the software. Unfortunately, they were little used in the Gulf operation. Partly this was because the staff had not practised them on peacetime exercises and had therefore not had a chance to evolve their own procedures based on these relatively recent AFSOPs. There was a quite understandable reluctance to change to new and unrehearsed procedures just before a battle. Additionally, some AFSOPs were not fully suited to the actual operational needs and, as the staff were unable to change any forms in the DICS system, they tended to reject those that were less than satisfactory. In some cases, in fact, as we discussed earlier, staff developed their own specific reports and returns in DICS using the word processor or spreadsheet facilities.

Those DICS AFSOPs that were employed were found to be somewhat pedestrian and laborious in use with far too many confirmatory keystrokes being required by the software, particularly when printing. This criticism highlights a major problem with Human Computer Interfaces (HCIs): there is a real need for an HCI to in some way adjust the nature and quantity of the information it provides and the format of the information it demands as the user evolves from being a novice to becoming an expert.

The DICS Audit Trail

In October 1990, Director CIS (Army) directed that an 'audit trail' be established for those activities that were specifically related to the DICS project and for those more general matters that concerned Op GRANBY Army CIS. An SO2, then employed with the Computer Assistance to Static Headquarters (CASH) team at Worthy Down, and subsequently reassigned to work under the Army CIS Agency (ACISA) at Blandford, was given this task.

The Historian in Residence

Although he initially saw his role as an 'historian in residence' concerned only with recording the events as they took place, he soon became a spotlight for the early detection and notification of problems. The records show that his very proper concerns, as expressed in his audit reports throughout this period were quickly acted upon by DCIS(A). There is little doubt that such a control mechanism helped to focus, during a busy and difficult period, attention and resources where they were most needed; it is an approach that is to be commended most strongly.

The Data Gathering Exercise

Shortly after the war was over, a DICS data gathering exercise was initiated as part of the audit trail with an SO1 collecting in person the views and observations of the staff in the Gulf. These were supported by detailed questionnaires which were also sent out to all the staff concerned.

The outcome of the exercise, when combined with the audit trail work, was a number of valuable summary reports that can now be used as the basis for establishing an MOD 'lessons learned' paper. Perhaps the most fitting assessment of the DICS project comes from the opening remarks of the main report:

> Whilst some may have criticised the detailed implementation of DICS there are few recipients of the system who have not accepted that the

net effect of DICS has been beneficial, in some cases dramatic. Most of the system's limitations stem directly from its late fielding during the critical lead up to the land battle and indeed during it, which are difficult times to develop and adopt new procedures. There is little doubt that DICS would have been a resounding success had its key facilities been provided and practised in peacetime before the Force was deployed to the Gulf.

Other CIS Support Systems

The Air Staff Management Aid (ASMA)

DICS was not, of course, the only computer-based system in the Division. The primary Force level strategic CIS system was ASMA with terminals in theatre at Main, Alternate and Rear Divisional Headquarters, as well as at the FMA, FFMA and BFME Force Headquarters. It provided the principal means of passing classified information between Joint Headquarters (JHQ) in the UK and the main theatre headquarters. At the start of the operation and before the Division deployed, ASMA had been issued to 7th Armoured Brigade, but it was subsequently withdrawn; 4th Brigade were never issued with the facility.

All the ASMA terminals in theatre were connected via satellite communication links to the central computer at High Wycombe. There were two

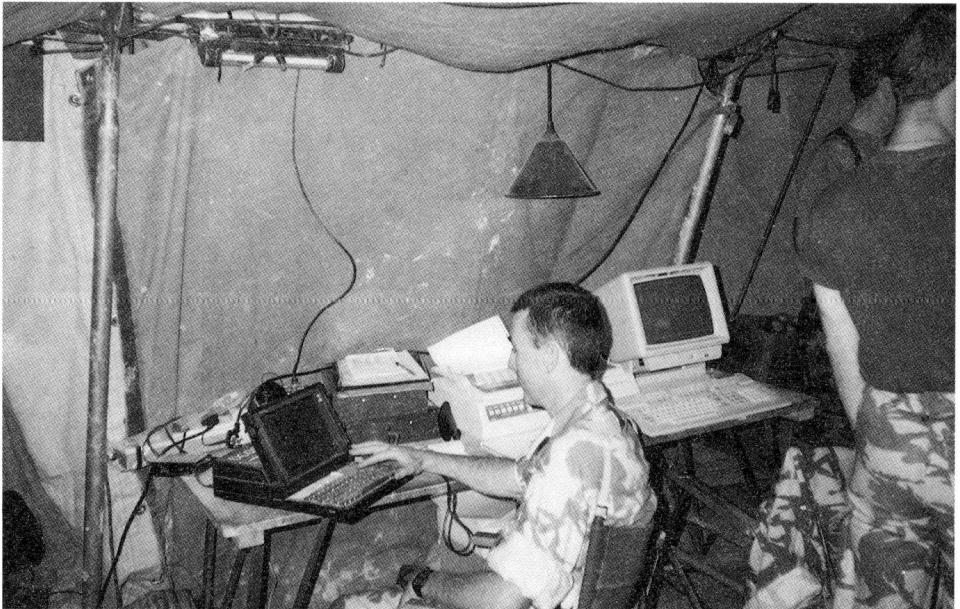

FIG 5.13 GRiD 1530 and Dismounted ASMA Terminal

ASMA terminals in Main Divisional Headquarters: one mounted inside the NBC vehicle for NBC reporting and the second dismounted as a common user facility, primarily for the Air and G2 staffs. The photograph at Figure 5.13 shows the dismounted ASMA terminal on the right of the 6-foot GS table with a staff officer using a GRiD 1530 computer.

ASMA had originally been developed by ICL as an RAF 'all-informed' tote board system. It was initially selected for Op GRANBY because, when the crisis started, RAF units were the first British deployment ashore in the Arabian peninsula. Subsequent decisions to provide major land forces in support of the Coalition changed this, and by the time the full extent of UK deployments became clear, the high level command and control structures had been put in place, ASMA had become the strategic command and control system, and JHQ High Wycombe had become the UK command head-quarters. In the event, its functional suitability for command and control and its user friendliness proved ASMA to be a wise choice.

ASMA operates using a series of tote 'pages' to which anyone with an ASMA terminal may gain access, although such access can be limited to specified totes. ASMA totes may be used to pass information between individual users, but that information is, of course, also on view to anyone else with access to that particular tote. Totes were used for a wide variety of purposes on Op GRANBY, some of which are outlined below:

- Problems with the provision of individual items were often resolved quickly by Supply staff in the Gulf dealing direct over ASMA with pro-visioners in the UK;
- The NBC cell at Divisional Rear used ASMA to get immediate answers to NBC problems from the Defence NBC Centre at Porton Down;
- ASMA general purpose totes were used constantly to pass news, weather updates and NBC information throughout the Force;
- *Scud* warnings were passed direct from JHQ to the Force over ASMA, providing the earliest possible pre-attack notification;
- Communications from the Force back to the UK to the Joint Headquarters and to the Ministry of Defence.

ASMA was found especially useful for rapid problem resolution both within theatre and between the Gulf and agencies in the UK, a summary of which is listed in Figure 5.14. Information transfer over ASMA was many times faster than normal signal messages, which could not in any case pro-vide the ASMA 'chat' facility between a number of different users.

There is little doubt that ASMA is a powerful and useful system for making state board information widely available to a community of users. However, it may cause problems when information of a discreet nature becomes widely viewed by a community who may not all have the necessary understanding or additional data to put it into proper context.

Some efforts were made to impose chain of command rules on the passage of information via ASMA. The logisticians established a Data Fusion Centre at HQ BFME whose sole purpose was to provide accurate and appropriately

- **Supply staff in the Gulf dealing directly with provisioners in the UK**

- **NBC cell at Divisional Rear getting immediate answers from Defence NBC Centre at Porton Down**

- **Passing of News, Weather and NBC updates throughout Force**

- **Scud warnings from JHQ disseminated throughout Force**

FIG 5.14 Some Uses of ASMA

filtered logistic information to JHQ. It is clear that an information facility that bypasses the normal chain of command can be of great value; it does, however need much thought about how it is to be used and careful development of appropriate SOPs if its not insignificant disadvantages are to be avoided.

Mapper

Much has already been said about Mapper, also known as the Command Support System (CSS), in the previous chapter. Like ASMA it relies on a central computer facility, which is based at Headquarters, United Kingdom Land Forces in Wilton, and had terminals connected into the Gulf via strategic communications links. Unlike ASMA, Mapper is not a tote system but primarily a collection of applications which are aimed at specific functions.

Movement planning, casualty reporting and messaging were amongst those most used in the Gulf, but because of its particular functional nature it had little visibility at Divisional level. It was used, however, to transfer Notification of Casualties (NOTICAS) from Field Records at Al Jubayl to the UK and to collect them from those field hospitals which had Mapper terminals.

Vigilant

Vigilant was an attempt to integrate ASMA and DICS, such that ASMA would carry DICS-type formats over its classified strategic circuits. The plan was that it would use DEUCE terminals in the Divisional Headquarters so that it had a 'smart' front end, providing DICS formats locally that were then manipulated within ASMA. Although this was staffed for approval and was intended to be used at Divisional Main and Rear, it never came about, much to the relief of the CIS staff officer, who was concerned about introducing yet another untried computer system into the headquarters.

BRUHN, GOLDWING and T-VITS

The new NBC reporting package known as BRUHN and purchased from the Danes has already been mentioned. It was originally intended to operate on a Sharp laptop. The Toshiba laptops purchased for it were found to have insufficient memory capacity and so some six Toshiba 3100s, purchased for DICS, were re-allocated by the CIS staff to the Divisional NBC cells. A photograph of the BRUHN terminal is shown at Figure 5.15. The system worked very well, enabling the effects of a ground-zero contamination to be rapidly determined. Most fortunately its services were never required in anger.

GOLDWING and T-VITS are both American MSDOS-based computer systems that were issued to the G2 DASC vehicle in the Division. GOLDWING is a text word-processing system which is linked by secure High Frequency (HF) radio. It was a cumbersome system based on a GRiD computer with Bernoulli drives and there were technical difficulties with this particular installation such that it rarely worked well. T-VITS, which also operated over secure HF, was supposed to provide photographic-quality video; this too had

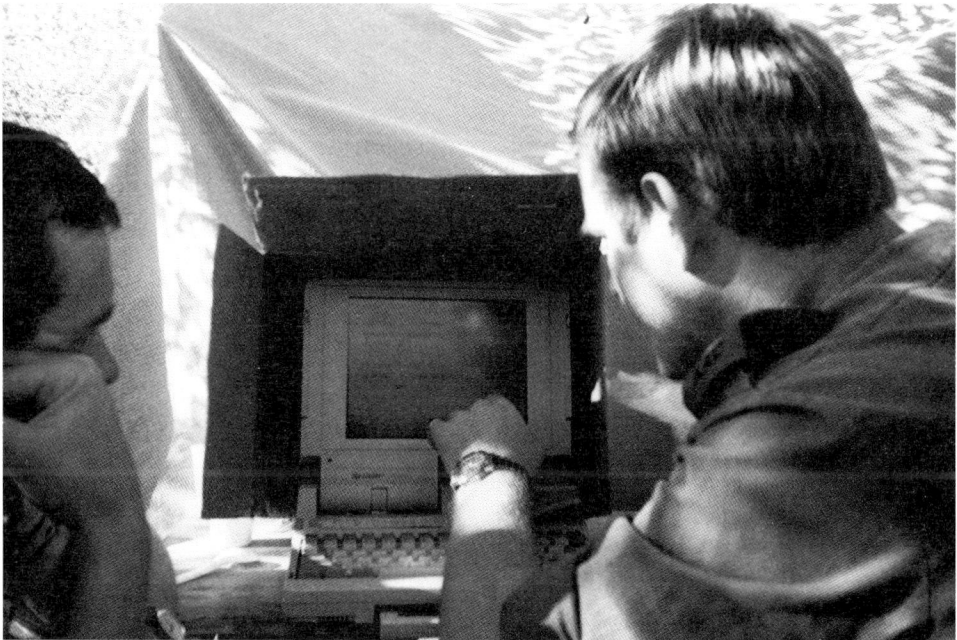

FIG 5.15 A BRUHN Terminal

technical problems with the particular installation such that it never functioned at all.

OLIVER, COFFER and Log Email

To complete the picture, mention should also be made of OLIVER, COFFER and Log Email. These are all logistic CIS facilities with OLIVER and COFFER being existing in-service systems, and Log Email a covert addition that came rather as a surprise to the communicators who provided the bearers. Again there was little involvement with these systems at Divisional level with just a single terminal off the theatre system at Rear.

We shall be looking at some of these systems again later; for now, the next part of the story takes us back to the beginning of the operation to see just how the complex network of communications systems, on which all the rest of the CIS depended, actually came about.

6.

How We Got There: III – The Communicators' Story

To actually extend communications over 1000 kilometres when we've been used to 120 in Germany is really rather difficult.
Commander CIS, Op GRANBY, June 1991

Introduction

Out of Area Operations

There are well established and well practised procedures for setting up joint service command and control facilities for OOA operations. Broadly speaking, lower priority was accorded to operations which might take place outside the central European theatre, reflecting the predominant threat there. This idea of priority was very important. It reflected, in terms of the equipment and manpower assigned, the commitment that was made from the Defence Votes to the particular activity. The OOA sphere came second to BAOR and Northern Ireland on the grounds that UK involvement in the former was likely to be relatively small scale and that such operations were less likely to present a critical threat to UK interests. The two most recent OOA operations (the Falklands and the Gulf) showed that the deployment of major forces could, nevertheless, be involved.

The philosophy of OOA communications is initially to establish what is known as a Forward Mounting Base to act as an operational and logistics springboard into the theatre of operations (see Figure 6.1). In the case of the Falklands this Forward Mounting Base was located on Ascension Island. Communication 'rear links' are then established from the Forward Mounting Base back to the United Kingdom. These rear links usually take the form of a satellite channel backed up by 'long haul' HF radio connected into the Defence Communications Network (DCN). The HF backup circuits can normally support no more than a simple 75-baud telegraph (Tg) message centre facility. A two-star headquarters may then be set up in the theatre of operations, and communications, again in the form of a satellite link backed up by

HF radio, are established from the port (or airport) of entry into the theatre back to the United Kingdom.

This philosophy has resulted in there only ever being a perceived need to earmark two satellite stations, with associated HF radio, for Out-of-Area operations. The equipment and manpower establishments for the associated command and control have thus always been based on this concept of operations.

OOA Communications

Since the Falklands War, the TRC 521 HF radio rear-link stations, most of which are held by 30th Signal Regiment, have been modernised and, at the time of the Gulf crisis, were considered by the communicators to be good, reliable systems which worked well. Apart from having to use a large, bulky generator, two of which would fill a *Hercules* aircraft, the equipment is of a high quality and provides the means for an excellent HF backup. The same could not be said, however, for the satellite terminals.

In mid-1990, 30th Signal Regiment had three TSC 502 Satellite Communications (Satcom) terminals of which they had only ever deployed two at the same time, needing to cannibalise from the third in order to keep the other two going. The technology of the 502s dates from the 1970s and by the mid-1980s their age and wide use was beginning to be reflected in declining reliability.

Unfortunately, not only were the 502s very unreliable, but in recent years there had also been a dramatic increase in the demand for data circuits for

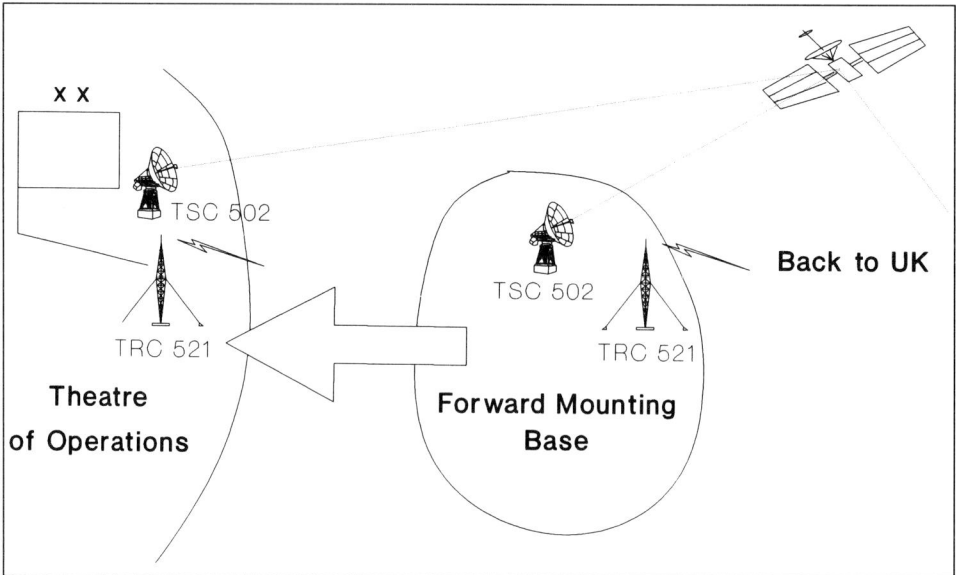

X X

TSC 502

TRC 521

Theatre
of Operations

TSC 502

TRC 521

Back to UK

Forward Mounting
Base

FIG 6.1 Forward Mounting Base with Two-Star HQ

command and control use over the satellite links on Out-of-Area operations. The three main command and control data systems: ASMA for air, OPCON for sea, and Mapper for land, were all now required. The TSC 502 satellite communications terminal could no longer cope.

Nevertheless, the staff, in practice, needed their command and control and logistic data systems, and thus the communicators were required to provide them on OOA operations. Without a Staff Requirement (SR), however, there could be no funding for additional facilities, so the existing equipments just had to be stretched with no immediate replacement in the offing.

The Start of the Operation

From the viewpoint of the Commander of 2 Signal Brigade, the GRANBY operation actually started at 4 pm on Wednesday, 8 August 1990. It was on that day that High Wycombe rang through to say that a single service RAF air operation was to be mounted in response to the Iraqi invasion of Kuwait on 2 August. Although it was to be single service, the RAF would much appreciate the communications support of 30th Signal Regiment to provide their rear links; was this possible? The Commander agreed and by 7 pm that evening the specific communications needs for this operation had been signalled from High Wycombe to 2 Signal Brigade and 30th Signal Regiment. They consisted of the three TSC 502 Satcom stations together with three TRC 521 HF rear links.

Initial Deployment of 30th Signal Regiment

At a co-ordination meeting the following day, Thursday, 9 August, 30th Signal Regiment were put on 72-hours notice to move to Saudi Arabia. By the time the representatives of the Regiment had arrived back in Blandford from that meeting, the order had been given for the Regiment's leading elements, a TSC 502 Satcom detachment and a TRC 521 HF rear link detachment, to be at Lyneham by 6 pm that evening. In the event, they did not get there until 6.20 pm; the *Hercules* aircraft, however, did not actually arrive until 8 pm, and, in the meantime, the detachments continued with painting their vehicles and equipments a more appropriate sand colour.

The two detachments were completely self contained with seven days' worth of rations and ammunition. They, together with their vehicles and equipments, filled three *Hercules* that night and arrived in Saudi Arabia on the Saturday morning, having staged through Cyprus.

That first move was a classic in how Army/RAF co-operation should work. Both sides knew the vehicles and equipments well, knew how they were to be prepared and airlifted, and knew what had to be done, by whom and by when. In terms of a joint CIS operation, it was an excellent example of a rapid and highly efficient Army/RAF deployment.

This first detachment went to the air field at Dhahran in eastern Saudi Arabia; an air field which was later to become the primary air base for the

RAF *Tornado* squadrons. The TSC 502 Satcom station was set up to work into a UK satellite earth station whilst the TRC 521 HF rear link operated into the DCN initially at Cyprus and then Gibraltar.

The Remaining TSC 502 Detachments

Over the next few days, two further deployments took place from 30th Signal Regiment with TSC 502 and TRC 521 detachments going to the air base at Thumrayt in Oman and a TSC 502 detachment going to the airport at Riyadh in Saudi Arabia. This latter detachment also began setting up a two-star headquarters in the city, for Riyadh was to host the Joint Forces Headquarters of the Air Commander, British Forces Arabian Peninsula. It was much later (1 October) that the headquarters was renamed British Forces Middle East (BFME) or 'Biffme' as it soon came to be christened. With the Satcom equipment resources of 30th Signal Regiment now exhausted, a further TSC 502 had to be borrowed from 264 Signal Squadron to deploy to the air base at Seeb in Oman.

As expected, the TSC 502s gave trouble initially, particularly from the effects of heat and dust, and there were immediate operational difficulties in finding suitable places to erect TRC 521 antennas on air fields. Eventually, however, these problems were overcome and by the middle of August, the satellite communications picture was as shown in Figure 6.2

The TSC 502 Capability

The circuits that could actually be carried by the TSC 502 were very limited. Originally, the equipments had been used in the Falklands to carry a 16-kbps speech channel together with a telegraph circuit. In the mid-1980s, however, channel access to the Atlantic DSCS satellite had been deliberately reduced from 16-kbps to 2.4-kbps, following significantly increased use of the satellite facility. The TSC 502 Satcom terminal had therefore been reconfigured accordingly to operate over a 2.4-kbps channel.

Since that time, practical communications experience from the series of joint service OOA exercises, referred to earlier, had resulted in the development of a revised configuration for utilising the available 2.4-kbps channel. At the time of the GRANBY Operation, the TSC 502s were configured to carry the circuits that are shown in Figure 6.3

In this arrangement, the 2.4-kbps channel from the satellite was split by a form of hybrid to feed two Code Division Multiple Access (CDMA) units each of which provided a 2.4-kbps speech channel and three 75-baud telegraph circuits. One of the 2.4-kbps speech channels was statistically multiplexed with three 2.4-kbps data circuits, and these were assigned to Mapper for the Army, ASMA for the RAF and OPCON for the Royal Navy. The second 2.4-kbps speech channel was used for the Defence Secure Speech System (DSSS or D-triple-S). In addition to possibly three 75-baud circuits that were assigned as common user (CU) telegraph, there were further 75-baud

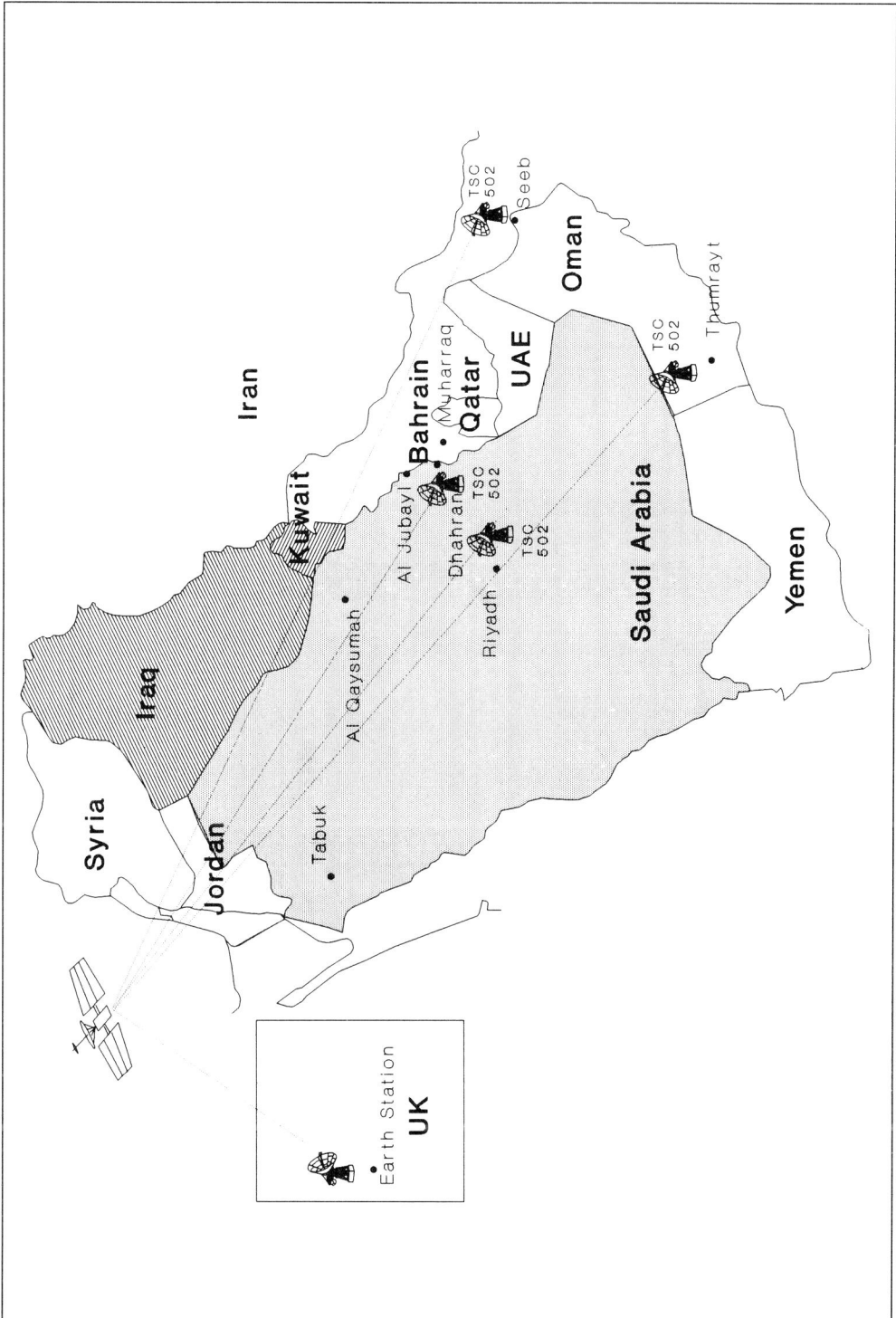

FIG 6.2 Initial Satcom Deployment, August 1990

telegraph circuits that could be earmarked for special purposes, such as tactical air request (TAR). Engineering was carried out over the engineering
order wire (EOW), and a number of encryption devices, referred to generally
as BIDs, were provided to give secure telegraph, speech and data, as
required.

The Operation Grows

By this time it was becoming apparent that what had initially begun as a
limited air and naval support task, in response to a request for help from the
Saudi government, was rapidly growing into a major, joint-service operation.
The political indications were that in addition to a possible air conflict, a
ground based task force would have to be sent to the Gulf. All this was going
to require rather more strategic command and control than just the four
communications detachments that could be provided by 30th Signal
Regiment and 264 Signal Squadron combined. Each of the air bases that was
going to be used in the Gulf needed to have its own satellite communications
links, as did the port of entry into the theatre at Al Jubayl and the force
headquarters at Riyadh.

Joint Headquarters and CIS Staffs

The Commander of 2 Signal Brigade had by this time despatched three
members of his staff, one of whom was his full colonel, to the main operations centre at High Wycombe. This centre had now taken on the role of Joint
Headquarters (JHQ) and here the three were to be instrumental in setting
up a joint CIS staff cell for Operation GRANBY. In addition, a Joint
Equipment and Manning cell was established, as part of JHQ, within
2 Signal Brigade itself. Consisting of five staff, this cell was responsible for
co-ordinating the issue of all the new command and control and communications equipment that was going to the Gulf. In the central staffs in the
MOD, the Communications Support Group (which perhaps more properly
should have been called the CIS Support Group) was also set up. This Group,
consisting of key members from all the relevant MOD branches, met twice a
week. It had the financial authority to approve and fund urgent new CIS
requirements for GRANBY. One of the first problems that these CIS staffs
now had to deal with was the Satcom terminal issue.

The New Satellite Systems

As we have suggested in earlier, luck once again was seen to play an important part in the CIS aspects of this operation. Although no high-capacity
satellite ground terminals could be bought off the shelf at this time, and no
manufacturer could produce suitable equipment in the required timescale, by
great good fortune, two major enhancements to the UK satellite communications capability were coming to fruition when they were most needed.

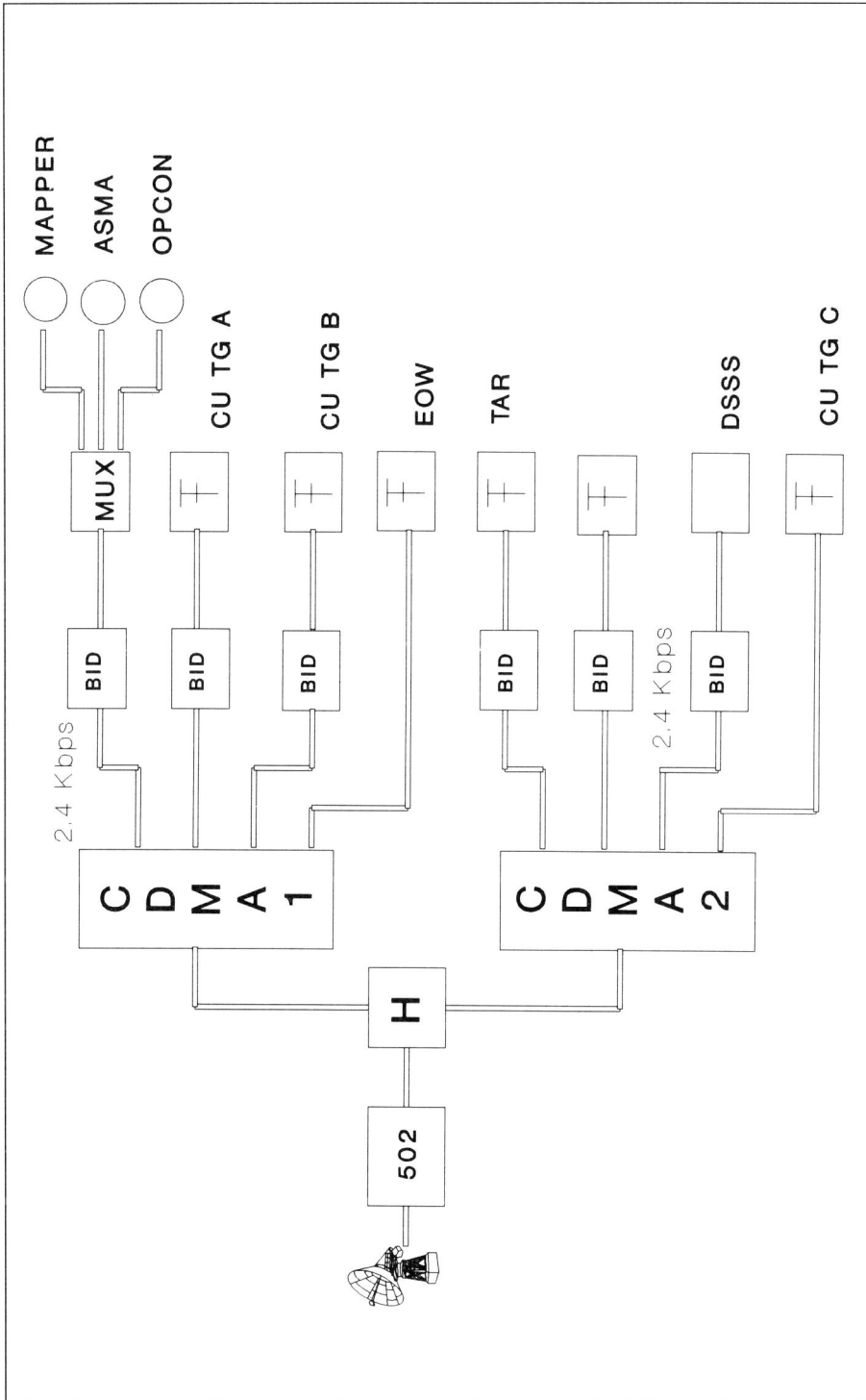

FIG 6.3 TSC 502 Configuration

As part of the *Skynet* 4 space segment programme, a third satellite in the series, *Skynet* 4C, was due to become operational in December 1990. That this existed at all was due to the foresight of an earlier Assistant Chief of the Defence Staff (CIS). All three *Skynet* 4 satellites were fully committed in the conflict, one having to be moved into a borrowed American parking orbit over Saudi Arabia. In addition, as we shall see later, the UK was able to trade some of the resulting *Skynet* 4 capacity for urgently needed US satellite ground stations.

For the second major enhancement, although there were no replacement Satcom terminals earmarked for OOA use, a new range of terminals for specialised BAOR use were just coming off the Racal production line and into service. These were the vehicle-borne VSC 501 stations. At the beginning of August there were only four of the new satellite stations actually in service, three of which were with the NATO AMFL Signal Squadron

The immediate answer to the communications problem was to divert the new VSC 501s as they came off the production line. Authority was granted by MOD and the VSC 501 systems were then sent straight down to 30th Signal Regiment at Blandford. Here, new detachment teams were rapidly formed up, given a smattering of Satcom training and then despatched out to the Gulf with the VSC 501 stations. As issued, the station was mounted in a single Land Rover; by the time it had completed commissioning at Blandford with its new crew and their equipment, it went out as a self contained detachment of six Land Rovers.

By late August the first of the VSC 501 satellite stations had deployed to the air base at Muharraq together with a TRC 521 HF rear-link station. By the outbreak of hostilities there were some 14 VSC 501 detachments deployed throughout the Gulf area. Once the VSC 501s started to become available, the elderly TSC 502 stations, which were proving so difficult to keep going in the harsh temperature and dusty conditions, were replaced and moved into back-up roles. Although the TSC 502 had many limitations, its modular design and the fact that it could be dismounted into buildings, still gave it some advantage over the VSC 501 vehicle-borne system, and it continued to play a useful part in the operation. The satellite communications picture by late October was as shown in Figure 6.4.

By this time, the *Jaguars* based at Thumrayt in Oman had been redeployed to Muharraq in Bahrain and a further *Tornado* air base had been established in western Saudi Arabia at Tabuk. The airfields at Tabuk, Muharraq and Dhahran each had two VSC 501 stations with a single TSC 502 remaining at Seeb with the maritime patrol aircraft. A VSC 501 was also deployed to Al Jubayl, together with one of the released TSC 502s and a further two VSC 501s were positioned at BFME in Riyadh, which also had another of the released TSC 502s. The fourth and final TSC 502 was then returned to 264 Signal Squadron.

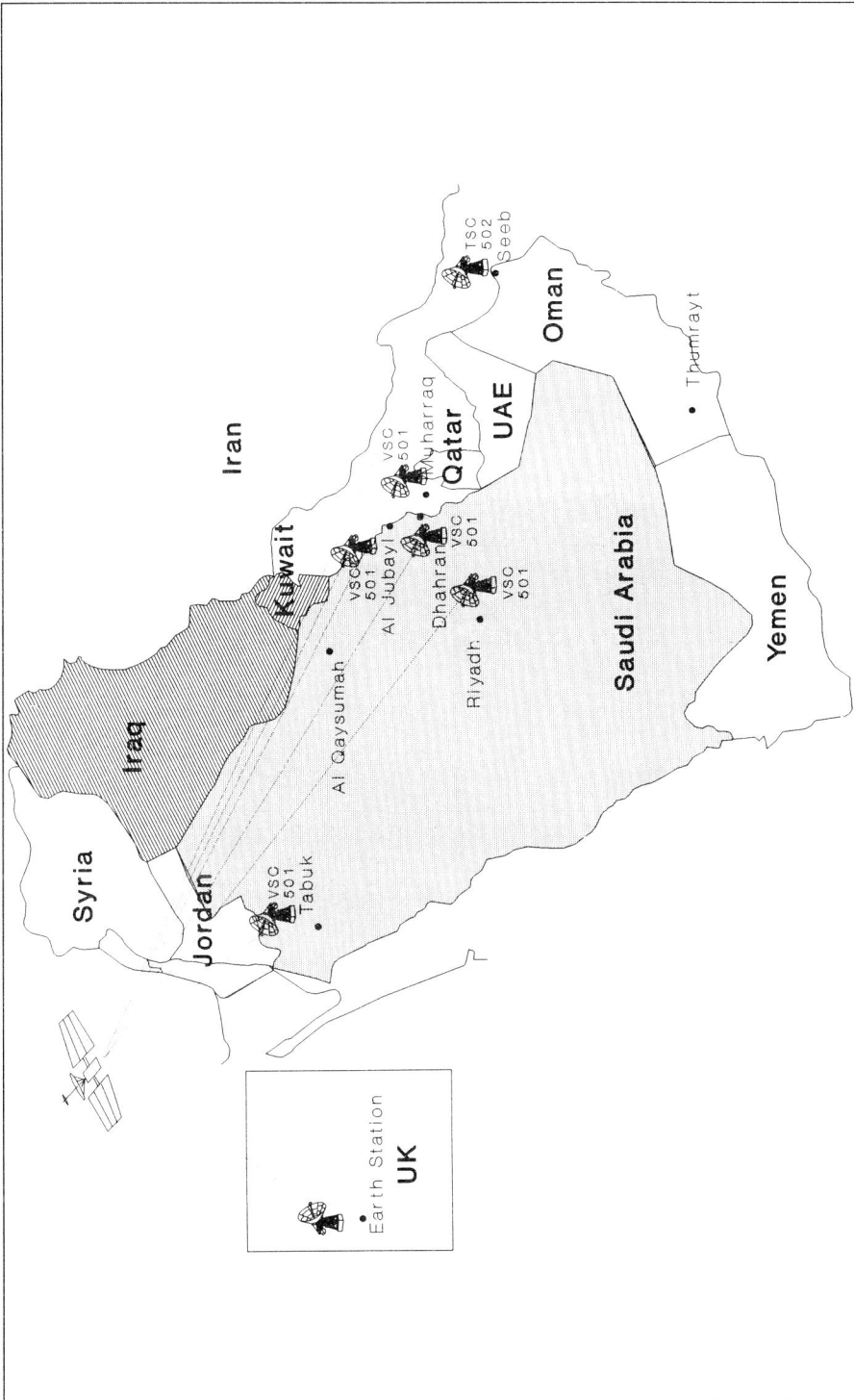

Fig 6.4 Deployment of VSC 501s

The VSC 501 Configuration

The VSC 501 station rapidly became the workhorse of the satellite communications system and it was configured as shown in Figure 6.5. The Racal 2636 modem provided a 16-kbps speech channel, a 2.4-kbps data channel, three telegraph (Tg) circuits and an engineering order wire (EOW). The 16-kbps speech channel was further exploited by the use of a software controlled GDC 16-kbps Minimux. This permitted the channel to be configured in a variety of ways up to a total aggregate of 16 kbps and was programmed, in this case, to provide six 2.4-kbps data streams.

Two further requirements were identified for transmission over the satellite bearer: that of a 64-kbps high resolution facsimile (fax) system for the targeting data used by the GR-1 *Tornados*, and provision of a more convenient, simpler to use secure speech facility in place of the DSSS. The new facility was called MENTOR and is discussed in a later section. It was carried over Satcom by expanding the capacity of the system to 85 kbps. This was achieved by the addition of a quadraphased shift-keyed (QPSK) modem, the output of which could be switched either to the high resolution fax or to a 64-kbps version of the GDC Minimux. This Minimux carried the three 16-kbps secure speech MENTOR circuits.

With all this equipment, the station, together with its data detachment and comcen, as mentioned earlier, comprised some six vehicles and 10 men.

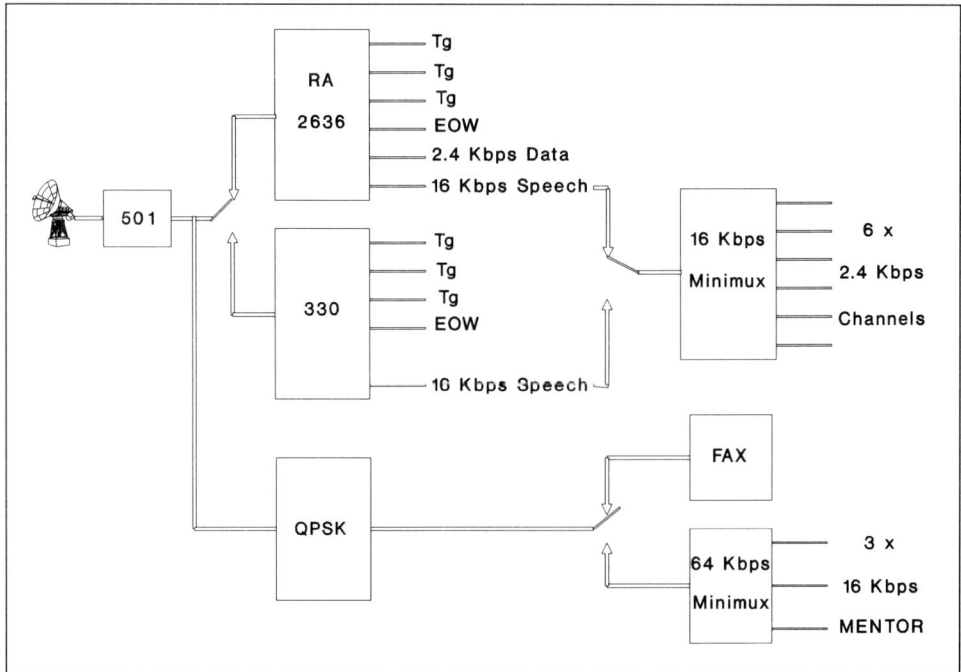

FIG 6.5 VSC 501 Configuration

This size was a distinct disadvantage, particularly since the equipment could not be dismounted from the vehicles. There were thus problems both in keeping it cool and in operating it in an urban environment. Nevertheless, the VSC 501 was, towards the end of the operation, successfully installed in an armoured vehicle.

Communications Facilities at the Air Bases

The two VSC 501s were typically configured, at each of the three main air bases of Dhahran, Tabuk and Muharraq, with one having the so-called 'GRANBY fit' and the other, the high-resolution fax for the *Tornados*. The standard 'GRANBY fit' gave access to the Air Staff Management Aid (ASMA), to the new secure speech system MENTOR, to DSSS, and to meteorological and common user 75-baud telegraph circuits. At Seeb, which only had a single TSC 502, with its lesser capacity, the new secure speech MENTOR facility could not be provided. In addition, where static landlines were available, these were used in preference the ASMA circuits which could then be run at a much faster rate. In such cases, the Satcom ASMA link acted as a back-up facility.

Another circuit carried by the Satcom which was found to be of great value was that of 'Tipoff'. This initially started out simply as a connection to a telephone which had a particularly loud bell but subsequently it was implemented as part of the ASMA attention getter (AG) system. It was connected through, on an all-informed basis, to a panic button in JHQ at High Wycombe where it was used to issue early *Scud* alert warnings throughout the force.

The vital importance of having a number of back-up communications systems is well illustrated by an incident that occurred at the Tabuk air base. The communicators there began having some problems with their local distribution system because of people driving over the cables. For the first month or so the problem did not materialise because the cables only gradually degraded. Then, of course, the telephone circuits began to fail at a most critical time, just towards the first week of the air war. It was therefore decided to dig the cables in.

A tri-national effort was established with the Americans from their air base providing the large mechanical trench diggers, the Saudis providing the plans to ensure that there were no utilities buried under the planned route and the UK Royal Engineers overseeing the project because they wanted a power cable dug in as well. All went famously until the power digger cut through a 200-pair, a 50-pair, a 25-pair and a 10-pair; all permanent civilian cables that nobody knew about but which were being used to provide the bulk of the landline communications within and out of the town of Tabuk. This cable break cut off the American sector completely.

The consequences of this incident could have been very serious indeed. The American air base depended to a large extent on landlines and it effectively became non-operational for the period it took the Royal Signals linemen, sitting in the trench for 18 hours, to repair the break. For the British air base, which had also lost all its landlines, rapid reconfiguration of circuits using the various alternative means available brought communications back to full operational status in just 40 minutes.

Secure Speech Systems

Another problem that was concentrating the minds of the CIS staffs was that of secure speech. At the start of the operation, the only secure speech facility that was available was the Defence Secure Speech System (DSSS). Although this had been in use throughout the MOD for some time, it was an old system with corresponding limitations. Anyone slightly hard of hearing has great difficulty with it and some of the older and more senior officers find it literally impossible to use. Another major disadvantage is that the telephone set is not positioned on the user's desk but has to be used from some common user booth, which is invariably a corridor or so away.

As Operation GRANBY developed so it soon became apparent that the staffs of the many diverse units and headquarters involved, both in the United Kingdom and BAOR, would have to carry out their detailed and often classified planning over the telephone from their desks if the time frame set for this operation was to be met. There was a requirement, therefore, for an alternative secure speech system to be designed and implemented as quickly as possible.

PATRON

The new system that was devised for the UK and for BAOR was called PATRON. This was based on the 218 System X main exchange in MOD, which acts as the hub of the network. All users are connected to this hub through their own digital exchanges by means of British Telecom 2-Mbps bearers, and these are bulk encrypted through a combination of various BID encryption devices. Access to these encrypted bearers is controlled by appropriate software programming of the 'subscriber's class of service profile' on the user's exchange. Within a relatively short time, the system was up and working.

With PATRON now providing a far better secure speech service within the UK and BAOR than the DSSS ever could, it was not long before the demand for secure speech into the theatre of operations became manifest. The need for staff to speak secure to HQ BFME at Riyadh or the FMA at Al Jubayl was undeniable and the only current options were DSSS or secure radio telegraph. A system had to be devised to carry PATRON into the Gulf. That system was known as MENTOR.

MENTOR

For MENTOR, three 16-kbps secure speech circuits were extended from the 218 MOD exchange into every location where a VSC 501 was deployed, via the main satellite dish at a UK earth station. At a much later stage, a further 12 16-Kbps secure speech MENTOR circuits were put into Riyadh via a TSC 100 satellite communications station that was brought in to a UK earth station. Again, to avoid misuse, all these circuits were routed via a manual operator switchboard in the MOD. An outline block diagram of the two systems is shown at Figure 6.6.

Distribution of Secure Speech within Theatre

Three approaches were adopted in distributing the secure speech facility from the VSC 501 stations within theatre. The simplest approach was to take the three 16-kbps secure speech circuits direct from the GDC Minimux, pass them through Rolli converters to correct the signalling, and provide each of three individual subscribers with a telephone on their desk. This was only suitable where the number of users was strictly limited and where they were all located within 400 metres of the satellite station.

The second approach was to use the EUROMUX system and extend the three MENTOR circuits into the subscriber interface unit (SIU). MENTOR could then be provided throughout the EUROMUX system as a secure, dialled, common-user facility for all subscribers. Local distribution via EUROMUX proved so successful that it was implemented at several major sites. The TRC 481 *Triffid* radio relay equipment was used to provide a 2-Mbps bearer circuit between buildings and key locations with appropriate BIDs for encryption and EUROMUX as the multiplexer and subscriber inter-face system.

Although the TRC 481 radio relay equipment is a 2-Mbps system and the EUROMUX is only a 512-kbps system, further exploitation of the

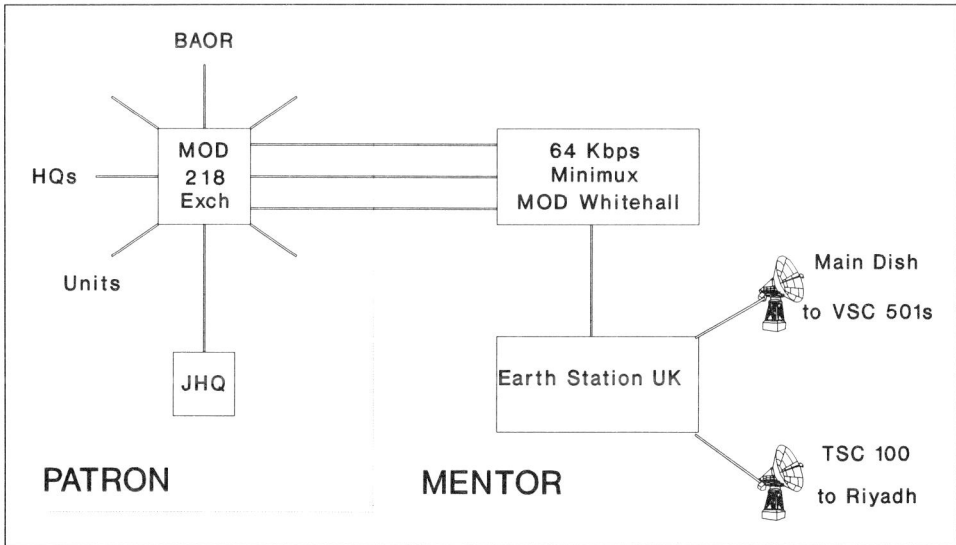

FIG 6.6 PATRON and MENTOR Secure Speech Systems

remaining 1.5 Mbps could not be achieved for lack of suitable multi-plexers.

The system was used widely and became known in Riyadh as the Riyadh EUROMUX Area Communications System or REACS. Similar facilities were known as TEACS at Tabuk, DEACS at Dhahran and so forth.

The diagram at Figure 6.7 shows some of the locations connected by REACS. The links are all provided by TRC 481 radio relay equipments, except for the quad cable from the RAF detachment at King Khalid International Airport to the hospital.

The third approach attempted for in-theatre distribution of secure speech, was to use the *Ptarmigan* system. The three MENTOR circuits from the Minimux were extended through Rolli converters via the Unit Line Control Equipment (ULCE) into the Unit Line Switchboard (ULS). This approach was not successful because of the poor quality of speech that resulted, mainly owing to accumulated losses throughout the system. It was also not popular with the staff because of the manual operator interface.

An outline block diagram of the three approaches is shown at Figure 6.8.

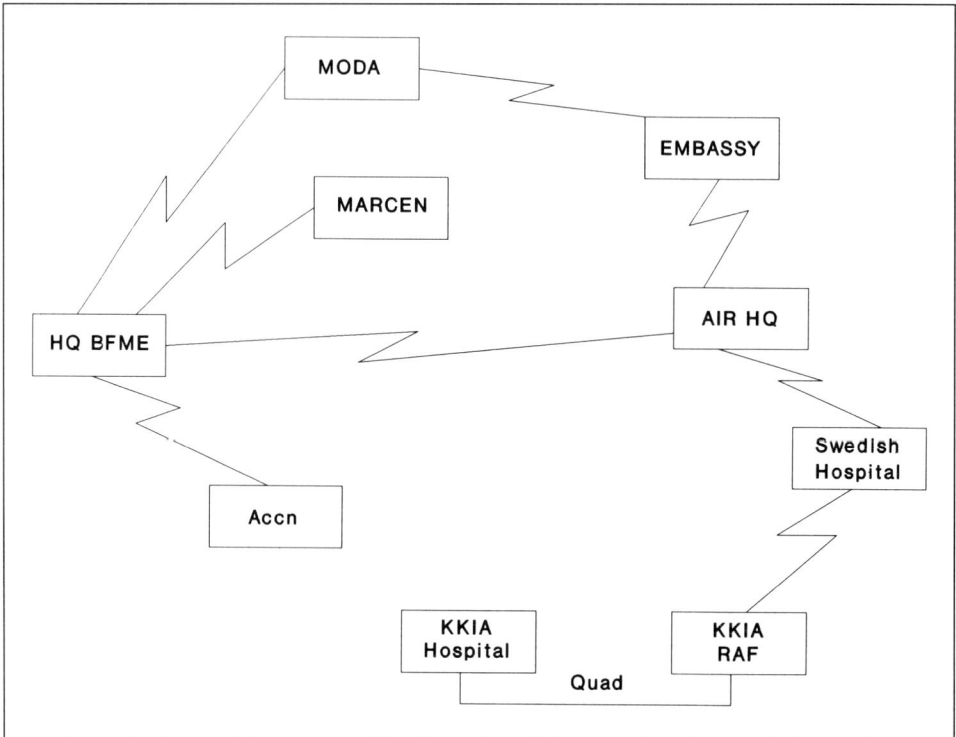

FIG 6.7 Riyadh EUROMUX Area Communications System

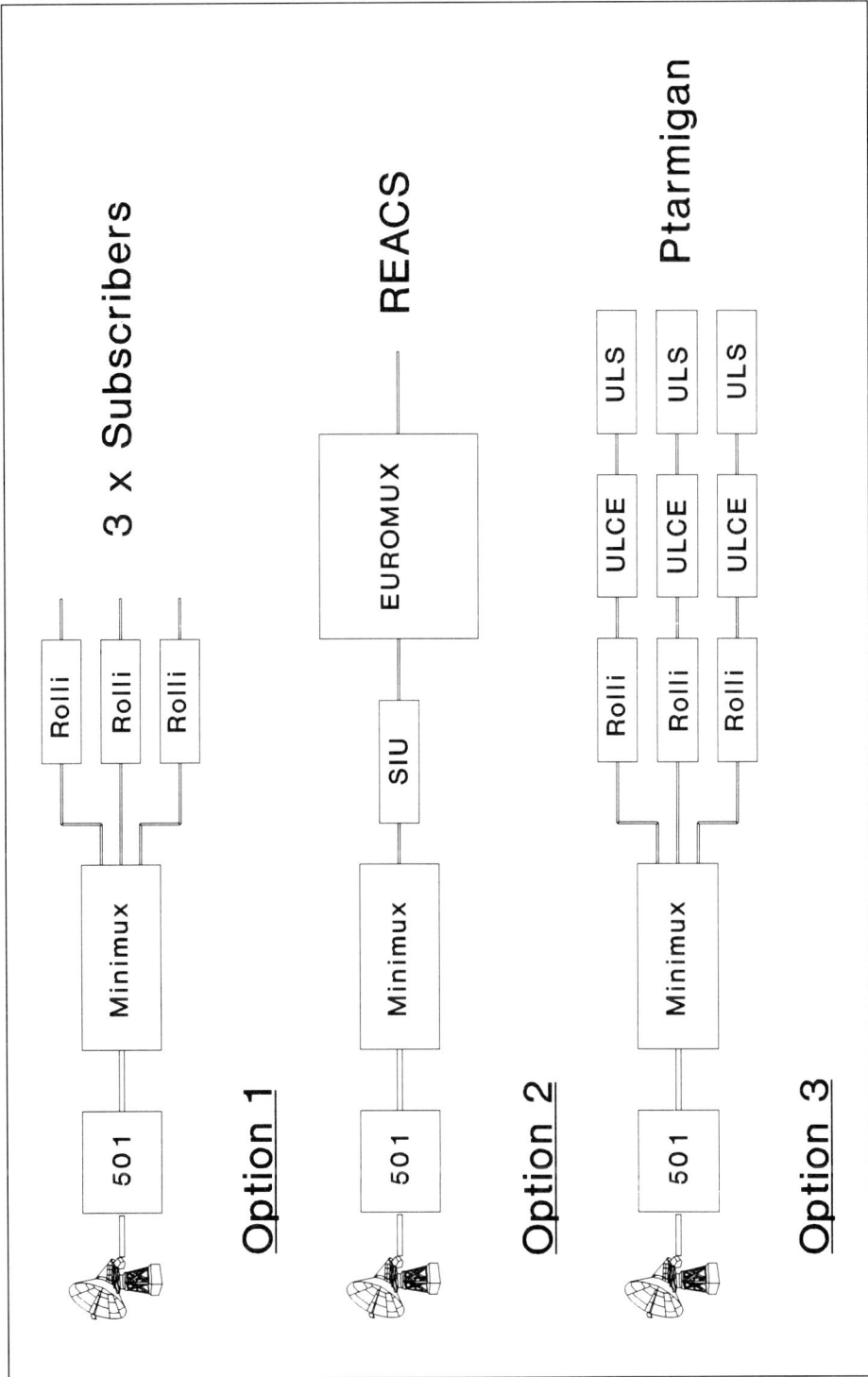

Fig 6.8 In-Theatre Distribution of Secure Speech

Advocate

The final development in secure speech systems for Operation GRANBY was a 64-Kbps extension to PATRON that was called ADVOCATE. This was required to permit secure speech between the Secretary of State in MOD, Air Chief Marshall Sir Patrick Hine at High Wycombe and Lieutenant-General Sir Peter de la Billière in Riyadh, at higher classifications than MENTOR could provide.

The ADVOCATE circuit was carried by the TSC 100 satellite station that is described later in the chapter. A proposal to provide secure video was turned down by the CIS staff on the grounds that there was insufficient bandwidth over the satellite channels to meet all higher priority operational commitments as it was.

A Clear Message

A message that became obvious to the CIS staff during the Gulf war, was that an effective secure speech system is an essential staff requirement in modern operations. It is needed at every stage; from peace, through transition to war, to war itself and it is needed between staffs in the UK, staffs in BAOR and staffs within the theatre of operations. The existing DSSS proved to be quite inadequate for this purpose.

The Data Explosion

At High Wycombe, the full colonel from the staff of 2 Signal Brigade who was heading the CIS cell involved with planning the command and control facilities was now faced with a major data-explosion problem. Everybody wanted his own particular command and control or logistic data circuit to be carried over satellite in the Gulf. There just was not sufficient capacity for them all.

We Shall Use ASMA

At the start of the operation, the colonel had agreed that ASMA should be used as the main operational command and control system. There were three reasons for this choice. First, the ASMA data streams were available direct from the mainframe computer which was co-located with the multiplexers in JHQ itself; this approach would thus avoid any problems associated with having to bring a number of different external mainframe data streams on to the system. Secondly, past joint exercises had shown that the protocols used by Mapper were almost unworkable over statistically multiplexed 2.4-kilobit circuits. This was because whenever Mapper seized the stream it tended to hog it and prevent ASMA or OPCON from gaining any access. In practice, ASMA, Mapper and OPCON were just not compatible with the TSC 502 configuration. Finally, it was felt that ASMA was far more user friendly than Mapper and was easier to engineer in the field.

Which Systems to Carry?

Now, however, there was a multitude of other information systems that particular staffs were demanding should be carried over the satellite. Initially, most of these systems were restricted to using landlines but these could not be relied upon. The quality tended to be poor with a best transmission rate of only 4.8 kbps. In addition, there were incidents such as a catastrophic failure of the international circuits in Egypt, which resulted in a total loss of all landlines out of Al Jubayl. The challenge now facing the CIS staff was to identify which of these information systems it was both operationally essential and technically feasible to carry by the satellite.

As a result, the CIS staff found themselves having to advise on and police operational priorities, an area which they felt was not their direct concern but more properly the province of the G3 operational staff. Nor were they becoming especially endeared to the staff users of those information systems that were not being granted satellite access. Despite the experience of Operation CORPORATE in the Falklands, there did not seem to be any SOPs or precedents to guide these CIS staffing processes. However, this criticism is not altogether fair, for although there had been a dramatic increase in the demand for CIS during CORPORATE no means had been found of meeting it.

Nevertheless, it would have been useful to have had the basic working SOPs in place, rather than the CIS staff having to develop them as they went along. Once again, the need to have in place a clear CIS management strategy was being demonstrated.

These were not the only problems. The majority of these information systems, such as OLIVER, used commercial hardware and were designed to work over dial-up modems using good quality civilian PTT lines. OLIVER, for example, simply would not work at Riyadh over Saudi PTT circuits. In the end it had to be carried by satellite because it was so central to the logistics activity. Even over the satellite there were difficulties with the commercial systems. The handshake times of their protocols did not account for the inherent delay over a satellite path, and the timing systems used varied from synchronous, through asynchronous to burstisynchronous. The timing problem was eventually solved, in the main, by ensuring that all data system communication links passed through the Minimuxes which were timed off the BT megastream as they passed through the UK earth station.

A Lesson to be Learned

The lesson to be learned here was that all procurement of systems for peacetime use should include consideration of their use over military bearer systems.

The engineering staff had some additional problems of their own. There was not sufficient space at the UK earth station to mount all the new Minimuxes so these were positioned at High Wycombe, many miles away.

The engineering order wires (EOWs), however, used to test and set up the circuits, terminated at the UK earth station. Engineering a data circuit that is terminated far away is no easy feat, and this problem was eventually overcome by extending one of the telegraph circuits from the field HQ end and placing teleprinters alongside both the data terminals. Engineering was then carried out between the field HQ and High Wycombe via telegraph.

It proved the point that systems need to have been planned and exercised in detail, end to end, before an operation makes them essential.

Ptarmigan

The main tactical communications network that was set up within the theatre was the *Ptarmigan* trunk system. Throughout October 1990, as 7th Armoured Brigade deployed to the Gulf, four *Ptarmigan* trunk nodes together with an OSC were established with the Brigade to provide it with its integral trunk and single channel radio access (SCRA) communications. At this stage the brigade was still under the operational control of the US Marine Expeditionary Force as it commenced its work-up training in the desert.

Shortly thereafter, the political decision was taken to increase the British ground forces to divisional strength and the Headquarters of 1st Division and 4th Brigade commenced their deployment to the Gulf over the Christmas and New Year period. This force, commanded by Major-General Rupert Smith was placed under the operational control of the US VII Corps, and with it came much of the rest of the *Ptarmigan* system from BAOR in the form of detachments from 7, 16 and 22 Signal Regiments, together with communications elements of 3rd Armoured Division.

As the strategy of the long left hook for the ground war was being developed, so 1st Armoured Division deployed further and further to the west, relentlessly stretching its logistic lines of communication to the FMA at Al Jubayl.

A Forward Force Maintenance Area (FFMA) then had to be established and a long line of *Ptarmigan* nodes, stretching for some 500 kilometres down the Tapline Road, had to be deployed, to link the FMA with the FFMA and the Division in the desert. Eventually, even the FFMA had to be split into two, FFMA(A) and FFMA(B), the latter being referred to in some reports as 'Area Keyes'. An outline of typical trunk communications connectivity at this stage is shown in Figure 6.9, with each inverted triangle representing a *Ptarmigan* trunk node.

In addition to these pressures on the logistic lines of communication to the FMA, there was also a problem in providing command and control communications between the divisional headquarters, far to the north and west, and HQ BFME in Riyadh. Secure speech was required across the tactical communications network between the force commander and the divisional commander and their respective staffs, and *Ptarmigan* could not be made to stretch any further. Indeed, the trunk links to the FMA

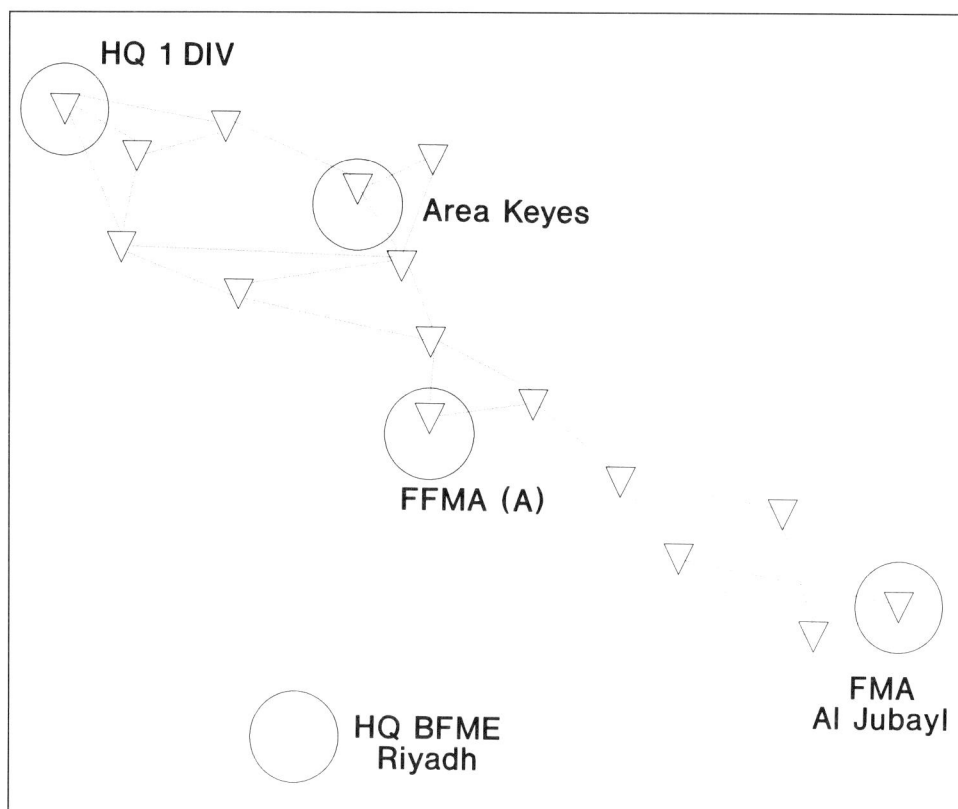

FIG 6.9 Typical *Ptarmigan* Connectivity

were almost certainly going to be broken when the Division moved into
Iraq. It is perhaps interesting to note that almost a corps' worth of trunk
communications assets were required to support an independent division
operating over large distances in a country with a limited communications
infrastructure.

The Ptarmigan Bridge

The most practical solution to these problems was to try to extend
Ptarmigan by means of a satellite link, something that had been discussed
by the communicators before but that had never been tried; to implement a
so-called '*Ptarmigan* bridge'. In order to meet the force commander's most
urgent requirement, the first Bridge would have to be between HQ BFME at
Riyadh and the FMA at Al Jubayl. This was to provide initially one and later
two *Ptarmigan* extended loop groups via the satellite thus giving access at
HQ BFME to the divisional trunk sub-system by means of the *Ptarmigan*
switch at the FMA.

A number of technical difficulties had to be overcome. Some brilliant work was done by RSRE Malvern in providing immediate working approaches to these technical problems. The software in the *Ptarmigan* switches had to be modified to accept the much longer time-outs associated with satellite working and a means for providing a coherent system of timing for the whole network had to devised. In fact, the timing system adopted provided only a partial solution and when it failed, the whole network had to be restarted from scratch. The operational need, however, was considered to be so great that this risk was accepted. A more complete and permanent solution is currently under development.

More Satcom Stations Needed

In addition, to these problems, further satellite ground stations, over and above the TSC 502s and VSC 501s already deployed, were going to have to be found. If *Ptarmigan* engineering control was to remain in theatre, and it was essential that it should, since this was where all the expertise resided, these new stations would have to have much larger dishes in order to avoid the necessity of routing the satellite links through the main UK earth station. With none available for off-the-shelf purchase, as we have already mentioned, the only possible source was the Americans. Here, the good fortune referred to earlier again came into its own. The CIS staffs were able to trade some of the *Skynet* 4 satellite capacity, which the Americans urgently needed (and which became available as a result of the launch of *Skynet* 4C), for a number of the big TSC 93 and TSC 100 in-service US Satcom stations.

Deploying the US Ground Stations

In late October, the first of these high-capacity satellite ground terminals, the TSC 93, arrived in the UK for trials and training. By late December it had deployed to the Gulf and in early January 1991 the first *Ptarmigan* bridge had been set up with TSC 93s at Riyadh and at Al Jubayl, as shown in Figure 6.10.

By the end of the month, the TSC 93 at Riyadh had been upgraded to a TSC 100 and a further TSC 100 had been positioned at a UK earth station. The second *Ptarmigan* bridge could now be established between Riyadh and the FFMA in the desert near Hafar Al Batin, thus releasing the long line of *Ptarmigan* nodes linking the FMA with the FFMA. These were urgently needed by the Division in preparation for its thrust into Iraq.

The TSC 100 at Riyadh, with its 20-ft dish, had sufficient capacity to act as the hub of an in-theatre satellite communications network linking together the FMA, the FFMA and a UK earth station with HQ BFME, as shown at Figure 6.11.

A third TSC 93 was placed with the division itself; this deployed tactically with an 8-ft dish, and because of satellite power considerations, it had to work to the main UK earth station rather than direct to Riyadh.

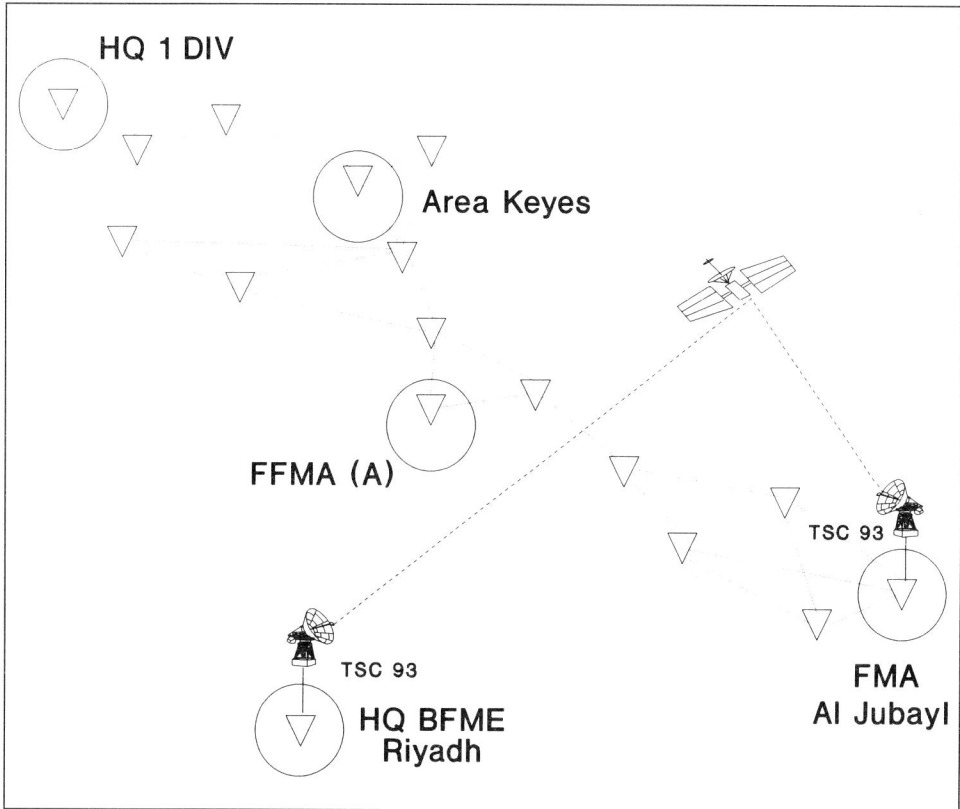

FIG 6.10 The First *Ptarmigan* Bridge

The TSC 93s and TSC 100s

As implied above, the satellite dishes for the TSC 93 and the TSC 100 Satcom stations could be either 8-ft or 20-ft in diameter. All but the TSC 93 that was deployed with the division, used 20-ft dishes. The basic difference between the TSC 93 and the TSC 100 station is in the number of 512-kbps ports that each has on its Tactical Satellite Signal Processor (TSSP). The TSC 93 has two ports and the TSC 100 has eight; hence the reason for the TSC 100 acting as the hub at Riyadh, although, in this case, only four ports were used. The TSSP ports were fed via Low Rate Multiplexers (LRM) each having 12 channels with an aggregate capacity of 256 kbps. However, the *Ptarmigan* 512-kbps signal could be passed straight into a port without the need for an LRM. Each channel of the LRM could be separately configured to carry either data or voice at bit rates up to 56 kbps and GDC Minimuxes were used to load each of these channels.

A block diagram of the TSC 100 station is shown at Figure 6.12.

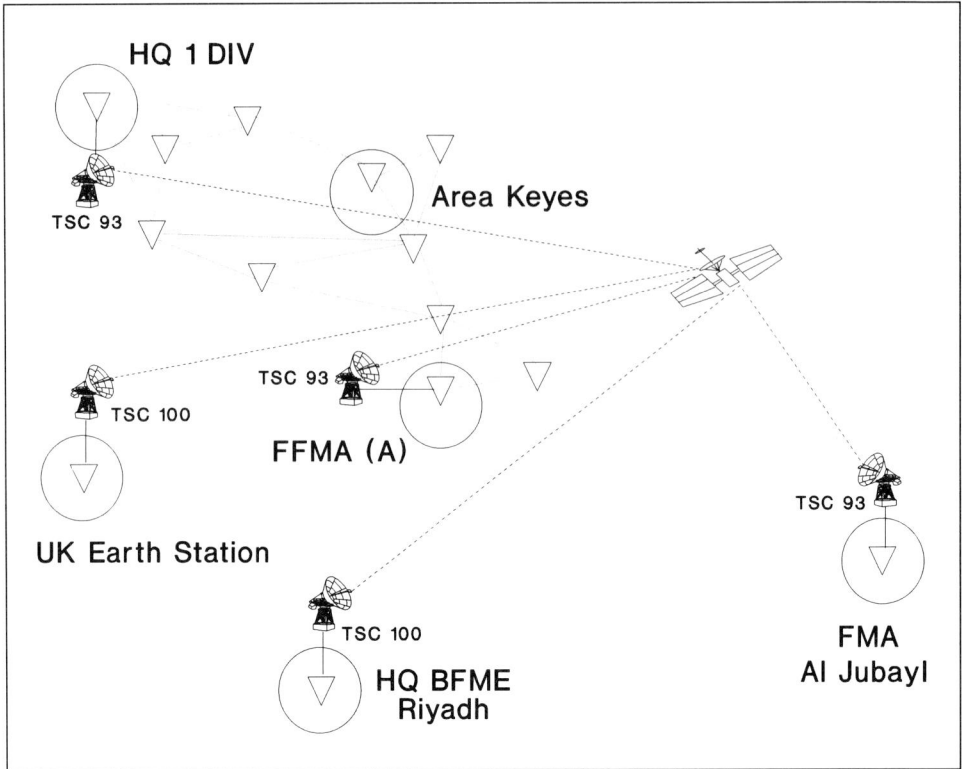

FIG 6.11 The Final Hub Network

Overall Communications Picture

Figure 6.13 gives an outline picture of the overall strategic and tactical communications that were in use in the Gulf, at the outbreak of the land war.

Satcom

All the VSC 501s and the single TSC 502 were linked to the main UK earth station, though these connections have not been shown on the diagram to avoid clutter. Each of the three main air bases at Dhahran, Muharraq and Tabuk is shown with two VSC 501 stations which carried the ASMA, Mapper, MENTOR, DSSS and high-resolution fax systems as well as a number of Tg circuits.

The air base at Seeb is shown with its single TSC 502, and, as a result, it lacked the MENTOR and high-resolution fax capabilities. VSC 501s are also shown at BFME, FMA, FFMA, and Divisional Main and Rear, and these provided each site with ASMA, Mapper, MENTOR, DSSS and Tg.

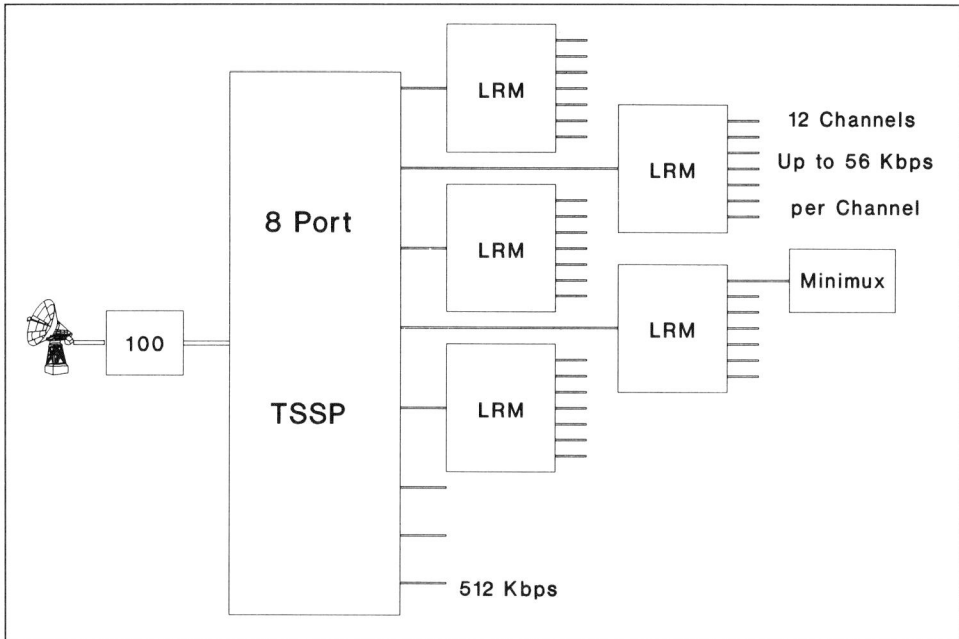

FIG 6.12 The TSC 100 Configuration

Ptarmigan

The TSC 100 hub system based at Riyadh is shown with its links to FMA, FFMA, Division and a UK earth station and these formed the *Ptarmigan* bridges that connected the four *Ptarmigan* segments together. Within the divisional area, static *Ptarmigan* was used to interconnect elements down to brigade level and SCRA gave *Ptarmigan* access to battle group headquarters.

A further refinement within the division was the use of *Ptarmigan* to switch the 16-kbps output from the VSC 501 to the duplicated QPSK and Minimux vehicles at either Main or Alternate, as required. This permitted the second VSC 501 to be deployed with HQ DAA, as we described in a previous chapter.

HF Back-up and Land Lines

Not shown on the diagram are the many HF rear links into the DCN that provided back-up communications and formal message traffic through the comcens. There were HF rear links at each of the air bases, at the FMA and at the FFMA. Land lines were also used at many locations, as back-up for those information systems being carried over the satellite, and as primary means for those information systems that had had to be turned down by the CIS staff. As has already been described, reliability of the PTT circuits in this part of the world is notoriously poor, and priorities had had to be assigned on the basis of operational need and technical feasibility.

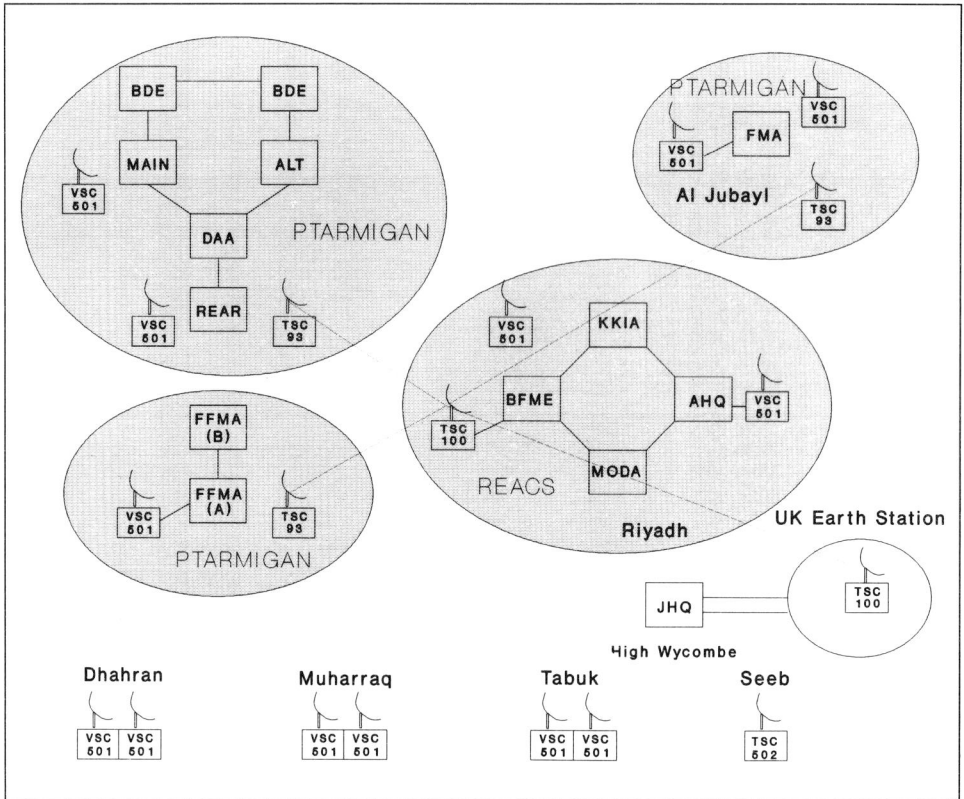

FIG 6.13 Overall Communications

Combat Net Radio and SCRA

Perhaps the most important aspect of communications, however, that has not so far been mentioned is that of Combat Net Radio (CNR). Once the ground war started it was effectively fought on Combat Net Radio with the fighting units using their integral *Clansman* VHF and HF radios almost exclusively. All orders were given via CNR. All the staff procedures that would normally be observed, such as the issuing of confirmatory orders and written orders, did not take place. Combat Net Radio orders were transmitted from Rear Division Headquarters direct to the Forward Combat Teams, at distances that sometimes extended beyond four hundred miles.

Severe communication problems were experienced both with the positioning and the number of the rebroadcast stations for CNR and with the positioning and affiliation of the *Ptarmigan* SCRA centrals as a result of these long distances and the very rapid advance made during the operation. Despite new doctrine to make best use of available assets in the advance, it was found impossible to keep up with *Challenger* and *Warrior* when the rebroadcast vehicles were mounted in FV 432. Similar operations will require new thinking if the communicators are to support such tactics effectively.

Commitments and Resources

Logistic Requirements

As with the communications assets, British land forces had to take with them to the Gulf more logistics assets than would have been required by a corps in north-west Europe; these were essential to support a reduced division for an offensive operation over the distances involved and given the particular requirements to move fuel and water. All these logistic units needed their command and control infrastructure to be in place in order to carry out their tasks. In BAOR and in the UK, most of these needs are met by static communications, provided by the local PTTs. In the Gulf the majority had to be provided by the use of military communications. Because the main *Ptarmigan* trunk system was deployed in the forward areas to gain the operational benefits of using ASMA, the hospitals and the logistic staffs were left with very limited facilities in the rear areas.

To perform effectively, logisticians need continuous visibility of their stock situation. One senior logistician compared, most unfavourably, the ability of Mr Sainsbury to determine from his database how many tins of baked beans he has and precisely where they all are, with the lack of real knowledge that he and his colleagues had about ammunition states once the convoys started down the Tapline Road from Al Jubayl. The equivalent technology was not in place, nor was there available the communications capacity to make it work.

One of the most useful systems used by the logisticians throughout the operation was known as Log Email and was briefly mentioned earlier. This was an unofficial system that was carried on the back of the official OLIVER system, without the knowledge of the communications staff, by using a form of sub-multiplexing.

It was not until after the operation was over that the communicators discovered by accident the existence of this system as they were closing down the operational circuits. A rerouted OLIVER circuit, by chance no longer had sufficient capacity for the sub-multiplexed Log Email facility and the staff users involved in the recovery phase complained of the loss to their signallers. They did get their Log Email system put back; but the incident did serve to demonstrate the importance of viewing CIS systems as a whole, rather than as the sometimes separated component parts of information system and communications bearer.

CIS Resources

At the start of Operation GRANBY, the total capacity of the existing equipment for Out of Area operations was a mere 2.4 kbps. By the end of the

Capacity in Megabits per second

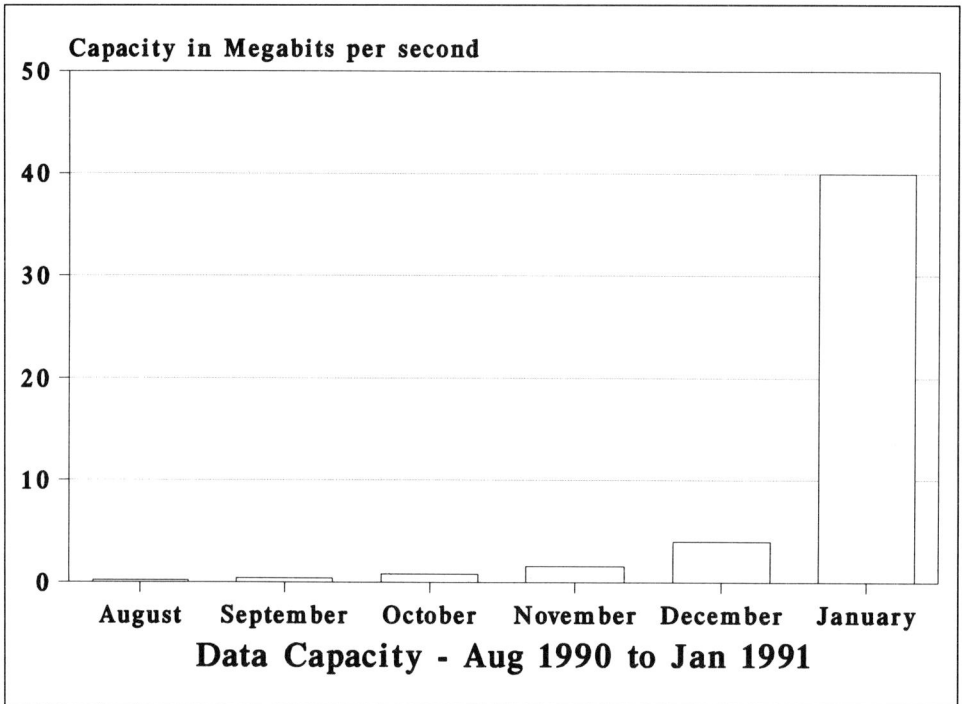

Data Capacity - Aug 1990 to Jan 1991

FIG 6.14 The Data Explosion

operation, the total capacity in use was over 40 Mbps and 25 different CIS information systems were being carried into the Gulf.

To provide this capability, nearly 3,000 Royal Signals personnel at peak were deployed in the theatre of operations. The rate of build-up of data capacity over the five months concerned is illustrated by Figure 6.14. Had hostilities commenced earlier, there would have been considerable problems with the command and control and communications systems.

7.

Assessing the Contribution and Lessons Learned

'That's the reason they're called lessons,' the Gryphon remarked: 'because they lessen from day to day.'

Lewis Carroll, 1865

The Contribution of CIS

By any standards, the military outcome of the Gulf War was an outstanding victory. The primary objective to eject Saddam Hussein's forces from Kuwait for the minimum loss of life on the Coalition side was achieved with emphatic success. It goes without saying that modern warfare is a joint effort, calling for the close co-ordination of all aspects of military power towards the achievement of a clearly defined goal. Despite the subsequent argument over what that goal should have been, there is no doubt that a major advantage on the Coalition side during the Gulf War was the relative simplicity and clarity of their purpose. Contrast that, for example, with the problems facing forces assigned to a UN 'peacekeeping' role in Bosnia. Co-ordination of effort, then, was the key to success. That co-ordination took place both within and between land, sea and air forces. Co-ordination of effort at this level demands a free and rapid exchange of information between the forces involved, within a framework of 'rules' which are understood and accepted by the commanders involved. There is no doubt that the 'men in the loop' are vitally important as the key, decision-making element of the system; but the capture, transmission, sorting and presentation of the vast amounts of information call for all the techniques and systems of CIS, applied in a responsive and flexible way in order to ensure the resilience of the overall command and control system.

In one of the most authoritative accounts from a British point of view to appear since the war,[1] General Sir Peter de la Billière recognises the joint nature of the conflict, and gives credit to the part played by the different elements of the force under his command and those of his allies. It is surprising to find that he makes almost no mention of the role played by CIS in

---

1 *Storm Command*, by General Sir Peter de la Billière, HarperCollins, 1992.

the Coalition success. To take an optimistic view, this could be a symptom of CIS's reaching maturity in Van Creveld's terms, that is, being taken for granted. However, in view of the extraordinary lengths gone to in order to provide the degree of CIS support finally available to commanders at all levels, this seems neither likely nor very fair.

From the evidence we have gathered from those who took part in the war, we have no doubt that whilst many features of the CIS provided must be regarded as having played a supporting role, there were others which, by themselves, must be considered to have been critical to the final outcome. Although outside the scope of this book, one element of CIS which certainly comes into the latter category is the computer support used to schedule the sorties during the air war, and to 'deconflict' the air space over the theatre of operations. To be fair, this is a contribution which is acknowledged in the de la Billière account of the war.

Very much within the scope of our book is the combination of *Ptarmigan*, Satcom and other tactical communications systems used to maintain contact throughout the UK force, from the Force Maintenance Area on the coast at Al Jubayl, all the way down to the individual units of the deployed force, both during the build-up and throughout the high-speed advance of the ground war. Without this communications infrastructure, it is hard to see how the commander's intention to transfer the UK force from working with the US Marine Division on the right flank, to being under command of US VII Corps out on the left flank, could have been achieved.

Ideal versus Actual

Having taken the approach earlier in the book to outline a theoretical CIS response to the situation in the Gulf, it is worth summarising the way in which the actual deployment compared with that theoretical assessment.

There were a few areas where the actual deployed systems far exceeded our theoretical assessment of what was necessary, but in the majority of cases the support provided fell short of the theoretical ideal. It is only fair to observe that in those areas where complex, highly-integrated CIS is called for and has not to date been successfully provided, it is unreasonable to expect the gap to be plugged at short notice in the run-up to a war. The CIS development which took place, particularly that of DICS, was an attempt to do as much as possible in the time available once the extent of the deployment was known and an assessment of capability gaps had been made.

Lessons Learned

As for the lessons learned, one may approach this on a number of different levels. First, a great deal of hard practical experience was gained in the use of CIS hardware and software under testing operational conditions. Many of these lessons could have been predicted in advance of the conflict, but there were also a few surprises. Going up a stage of abstraction, we must deal with the experience gained at the total system level and deal with the contribution made by CIS, within the scope of our book, to the 'business' objectives of the war. The third area where we must draw some conclusions is, so far as the future of CIS is concerned, probably the most important of all; this concerns the whole area of the specification, procurement, development and support of CIS.

We have taken a bottom-up approach in this final chapter and started with the low-level lessons derived from the performance of the hardware and software. The most important lessons concerning the way in which the systems were procured and the organisational significance of the way things happened are left until last.

Hardware

It was to be expected that the physical environment, particularly the heat, dust and sand, would take its toll on electronic equipment. As might have been predicted, the most vulnerable components were floppy disk (diskette) drives. Users became accustomed to having to take extreme care of both disks and drives, but even so there were data casualties. The problem was accentuated by the fact that, in the DICS PC-based system, diskettes had become a user-controlled item; in most previous battlefield systems (particularly *Wavell*), magnetic data storage media were the concern of system operators and managers.

A surprising success story was the durability of the one laser printer 'acquired' by HQ 1 Armoured Division in the early days of the build-up in Saudi Arabia, and which accompanied Main Divisional HQ throughout the war, that is, it was transported from one Main HQ to the next on change of command. Laser printers have quite a lot of moving parts, but they are well made and it is difficult for dust and dirt to get at the vulnerable places. The print engine at the heart of the machine uses the same technology as a photocopier and is inherently robust.

Differing power requirements were at best an irritant, and at worst a serious problem. Equipments requiring supply voltages of 240, 110, 24 and 12 volts were all used. This created confusion, the potential for damage and logistic problems, not to mention the size and weight of the convertors which had to be used. The clear lesson here is standardisation.

Another problem which could have been predicted was the reaction of users, accustomed to off-the-shelf commercial equipment, to the size and weight of near MILSPEC hardware. When users can see that the same functionality is available in a package much smaller in size and weight than the military version they are, not surprisingly, unhappy. This relates to a procurement lesson, set out later.

Software

The question might well be asked 'Why was Unix not used as the operating system for DICS?' As explained in Chapter 3, the original intention was to use Unix, since that was what was used by CORMIS. However, the majority of the most popular applications were MSDOS-based, and there was a shortage of the necessary Unix system management skills. In a system which relied very heavily on end users to manage as well as use it, it made sense to stick with an operating system with which they had some familiarity, even if it was less than ideal in some other respects.

An integrated package, UNIPLEX, formed part of the DICS software release, but almost no use was made of the integrated office environment features of UNIPLEX. This was probably a question of training and culture. There was no widespread previous experience of integrated packages, and the beginning of a war is not the time to learn new ways of doing things. This point is returned to under organisational and human factors.

Word Processing was the main application, and one which had a direct impact on the success of the operation, via the preparation and dissemination of orders.

Spreadsheets were used extensively. Supercalc was used much more than UNIPLEX, because in the main it was the package the users already knew. The special purpose spreadsheets developed by the DICS team were generally not used. There were many examples of good uses of spreadsheets in both operational and logistics sectors.

Databases: the INFORMIX databases developed by the DICS team, were also little used. Some 'home-grown' INFORMIX cardfile databases were found to be of practical use and PC File was specifically introduced to better meet the need for such databases.

Electronic Data Transfer: the DICS Email File Transfer over *Ptarmigan* became routine especially for Op Os and Frag Os. This 'transformed DICS from a useful word processor and calculator into a powerful and battle-winning CCIS'.

Graphics: no pointing device was available on DICS so the capability to draw was limited, but still used in order to create map overlays and graphical task organisations which could be sent by Email.

Reports and Returns: 130 formats based on AFSOPs were coded-up by the DICS team but not used. It is possible that the AFSOPs were too new to be readily adopted.

Procurement

The difference in procurement procedures between 'operational' and 'non-operational' is a hangover of organisational history which should have been cured long ago. In many areas the distinction is not clear cut or individual systems are used across the divide. Procurement can be correspondingly confusing with uncertainty about which of the distinct procurement routes to follow or with a notional requirement to follow both for different aspects of the requirement.

As the speed of procurement of EUCE and DEUCE showed, there was no need to drop competition just because there was a war on. A more fundamental issue was that many were unhappy with the systems procured. In part this was the result of procuring systems to a military specification, in particular as regards hardening, which meant that they were more cumbersome and therefore less mobile than some of the lightweight commercial systems available. In part it was the result of the short procurement timescale, which meant that neither training nor software could be brought to a level where users could take advantage of the considerable capability available in the machines procured; this included, for example, the fact that much greater use might have been made of the capability to network the systems with clear benefits for the conduct of the operation. For the future, there must be a place both for systems offering sophisticated capabilities including the facility for networking, and for truly mobile laptops offering the basic range of functions which will satisfy the needs of many staff officers in the field.

Organisational/Human Factors

In a sense, DICS and other support introduced to the Gulf was a step ahead of any support which had ever been provided in peacetime. With the exception of pockets of experience at BAOR (Officepower on CALAIS), 1 (BR) Corps (Uniplex on CORMIS) and 3rd Armoured Division (WP, Spreadsheets and Graphics on GRiD) there was little experience of using CIS/IT as a comprehensive support tool in a way which was integrated into working practices and procedures. Such a state lay in the (immediate) future of Army CIS, as envisaged in the CHOTS, CASH and UNICOM concepts and projects. Many staff officers and commanders were aware of the potential power of CIS/IT, but, for reasons of lack of training or limited requirement, few used the full capability of the systems available.

It is still seen as a novelty – but worse than that, it is a dynamic and rapidly changing novelty. In our earlier book we quoted from Martin Van Creveld on the adoption of new technology – one cannot consider the technology mature (and hence have any confidence that its potential is being adequately exploited) until the basic rules governing its employment and effects 'are no longer the subject of debate'. CIS capabilities will not stand still long enough to allow us the luxury of a long-drawn out assimilation process.

Much has been made of the idea that one should not change tried-and-tested procedures on the eve of a war. This has been used to account for the lack of enthusiasm for headquarters LANs or for the adoption of reports and returns based on AFSOPs. Whilst one has a degree of sympathy with this position, it tends to deny the fact that war has traditionally provided the stimulus for change of the most radical kind in equipment, procedures and tactics.

Perhaps this point illustrates another way in which the nature of war has changed in the last half-century. During the six years of the Second World

War, tactics, techniques and equipment evolved constantly in the search for advantage and in response to the other side's developments. The pace of modern, technology-based warfare is such that there is much less time now for such evolution.

Interoperability

Interoperability, or the lack of it, has been cited as the cause of numerous areas of difficulty. The problem of interoperability is well understood and steps are being taken to solve it, starting with the Army IS strategy and the identification of common standards. At the time of the Gulf War, most of the interoperability measures had not been taken.

One example described in some detail was PAMPAS, the unit personnel system, which can only interoperate with the mainframe personnel system at the RAPC Computer Centre. A serious 'fudge' was required to make it work with the Field Records System. To be fair, it was only ever designed to work to the RAPC Computer Centre, and there was no general purpose, flexible way of making systems interoperate with each other such as exists today in the form of the Open Systems Interconnection (OSI) architecture.

Communications

One of the most significant features of the communications effort was the sheer volume of communications provided. OOA contingency planning was based on a much smaller force with two Satcom stations and HF back-up, one for the Forward Mounting Base and one for the Joint Force Head-quarters. The actual deployment, as we saw in Chapter 6, involved a complex network of Satcom and other communications means with a total capacity hugely in excess of the OOA contingency plan figure.

The inherent flexibility of modern, digital telecommunications systems was demonstrated time and time again as the network and its utilisation for all forms of traffic grew during the deployment. There is no doubt that the future of military telecommunications lies in the adoption of standards developed for the civil field, which is a complete turnaround since the 1950s and 1960s when systems like *Ptarmigan* were first being designed. The use of civil standards not only cuts procurement costs, but greatly increases the flexibility with which military systems can be used in conjunction with existing civil capacity to respond quickly to new situations.

Rapid Development of Capabilities Proven Technically in Peacetime

If, as suggested earlier, the pace of modern warfare makes the sort of evolution of technical capabilities one saw in World War II no longer possible, there is one way in which capabilities can be introduced rapidly when needed and which relies on a similar principle. This is by the tracking during peacetime of developments both in technology in general and in their application by potential enemies in particular, and by carrying out the necessary research and development to improve one's own potential capabilities, without spending large sums of money on full production and deployment. Solutions can then be 'taken off the shelf' when needed and put into production. There are, of course, many limitations; for example, the required production quantities could be impracticable in the time available, and the R&D may not have adequately covered all the practical aspects involved in deploying the solution. However, there are some good examples of the application of this approach in the Gulf War.

The Satcom adaptor for *Ptarmigan* had been worked on at the Royal Signals and Radar Establishment (RSRE) during the 1980s and was a more or less proven technique, placed 'on the shelf' for the lack of funds to deploy it fully. Some further development was needed to overcome the problem of system timing – time-outs designed into the system protocols could not cope without modification with the relatively long time taken for a signal to travel up to a geostationary satellite and back to earth. The problems were solved and a workable solution produced by the RSRE during the Gulf War build-up, and the use of the adaptor made a major contribution to the flexibility and mobility of the UK force.

A secure speech adaptor for the PRC 351/2 manpack VHF radio transceiver had been developed in the late 1980s as part of a programme to improve the capabilities of the *Clansman* range of radios, but was also shelved for lack of funds. In response to an urgent request from the Gulf, the design was taken off the shelf, and the required number of adaptor kits were made by hand and tested by a team of Royal Signals technicians and RSRE scientific staff. The adaptors were flown out to the Gulf and installed *in situ* under appalling working conditions, almost on the eve of war.

The Gulf War also gave an added impetus to the replacement of DSSS by the more convenient, simpler to use PATRON secure speech system.

The examples given above are all of discrete items of equipment which were needed to fit into an established system in order to improve its capabilities, and in those cases the concept of 'contingency R&D' seemed to work well. The concept clearly would not work when the 'contingency solution' is not an item of equipment but a system whose introduction would require changes in procedures and a degree of training.

The Legacy of Operation GRANBY

It seems inconceivable that an operation of the size and scope of the Gulf War could become overshadowed by other events in such a short space of time, but such is the pace of geopolitical change in the early 1990s. At the start of the War, the process of reshaping the British Armed Forces in response to the fall of the Communist regimes of eastern Europe had already begun. It was of great concern to those out in the desert that the *Options for Change* programme was under way, creating uncertainty over the shape and size of the Army to which they would return after the conflict. At any time during the Cold War period, the idea of denuding 1 (BR) Corps of most of its capability, to fight a war in the Middle East, would have been unthinkable, yet there can be few observers who saw the removal of the forces from Germany as anything other than an accelerated part of a process which was going to happen anyway before too much longer. The subsequent deployment of British troops to protect the Kurds in Operation HAVEN, and their deployment to join the UN forces in the former Yugoslavia gave an indication of the very different and much more wide-ranging nature of the use of armed forces in the post-Cold- and Gulf War period.

An immediate legacy of GRANBY is the large amount of hardware procured, especially for DICS. This has now mostly been redeployed to form the nucleus of a 'permanent' headquarters and unit level operational CIS tool, as a step on the way to the fully-integrated Army CIS envisaged by the Army IS strategy, and being implemented by projects such as UNICOM, CASH and BICS.

On the human front, the operation has produced a body of people with first-hand experience, as commanders and staff officers, of the value of modern CIS support on the battlefield, although it is sad to report a prevailing attitude that 'experience of 100 hours of largely unopposed operations proves nothing'. Those who survive the redundancy programme under *Options for Change* ought to be in a good position to see that the lessons of the Gulf War are heeded by those responsible for developing and procuring CIS for the future, however uncertain that might be.

List of Abbreviations

1 MARDIV	US 1st Marine Division
4GL	4th Generation Language
9 ID	US 9th Infantry Division
ACISA	Army CIS Agency
Admin Ops	Administration Operations
AFSOPs	Army Formation Standing Operational Procedures
Arty Int	Artillery Intelligence
ASMA	Air Staff Management Aid
BAOR	British Army of the Rhine
BATES	Battlefield Target Engagement System
BICS	Battlefield Information and Control System
CALAIS	Command and Logistics Information System
CASH	Computer Assistance to Static Headquarters
CDMA	Code Division Multiple Access
CIS	Command, Control, Communications and Information Systems
CHOTS	Corporate Headquarters Office Technology System. Ministry of Defence office management system
CNR	Combat Net Radio
Comms Ops	Communications Operations
CORMIS	Corps Headquarters Management Information System
CRA	Commander Royal Artillery
CRE	Commander Royal Engineers
CRT	Cathode Ray Tube
CSS	Command Support System (Mapper)
DAA	Divisional Administrative Area
DADPTC	Defence ADP Training Centre
DASC	Divisional All Sources Cell
DCIS(A)	Director of CIS (Army)
DCN	Defence Communications Network
DEACS	Dhahran EUROMUX Area Communications System
DEUCE	Downsized End User Computing Equipment
DGITS	Directorate General Information Technology Systems
DGMS	Directorate General of Military Survey
DHICS	Desert HICS
DICS	Desert Interim CIS System

DMCP	Director of Military Communications Projects
DR	Despatch rider
DSSS	Defence Secure Speech System
Email	Electronic mail
EMC	Electromagnetic compatibility
ENG	Electronic news gathering
EOW	Engineering order wire
EUCE	End User Computing Equipment
FACE	Field Artillery Computing Equipment
FDC	Fire Direction Centre
FFMA	Forward Force Maintenance Area
FMA	Force Maintenance Area
FragOs	Fragmented Orders
G1/G4 LO	G1 Personnel and G4 Logistics Liaison Officer
G3 Ops	G3 Operations
GIS	Geographical Information Systems
GS	General service
HCI	Human Computer Interface
HF	High Frequency
HICS	Headquarters Interim CIS System
HQ BFME	Headquarters British Forces Middle East
Intsum	Intelligence Summary
IT	Information technology
ITT	Invitation to Tender
JCIS	Joint CIS
JHQ	Joint Headquarters
kbps	Kilobits per second
kbyte	Kilobyte
LAN	Local Area Network
LCD	Liquid crystal display
LO	Liaison Officer
Log Email	Logistic electronic mail
LRM	Low Rate Multiplexer
LTCs	Long Term Costings
Mapper	Also known as CSS, the Command Support System for Headquarters United Kingdom Land Forces
Mbps	Megabits per second
Mbyte	Megabyte
Mil Spec	Military specification
modem	Modulator Demodulator
MSDOS	Microsoft Disk Operating System
NBC	Nuclear, Biological and Chemical
NOTICAS	Notification of Casualties
OA	Office automation
OOA	Out of Area
OpOs	Operational Orders

PAMPAS	Personal Administrative Microcomputor Pilot ADP System. Unit orderly room pay and records system
PC	Personal computer
PCB	Printed circuit board
PD/MCS	Programme Director Military Communications Systems
PE	Procurement Executive
PERT	Programme Evaluation and Review Technique
PRIs	Presidents of the Regimental Institutes
PTT	National Postal, Telephone and Telegraph administration
QA	Quality assurance
QPSK	Quadraphased shift keyed
R2	Reports and Returns
RAM	Random access memory
RDBMS	Relational DataBase Management System
RDC	Receipt and Despatch Centre
REACS	Riyadh EUROMUX Area Communications System
RSRE	Royal Signals and Radar Establishment, Malvern. Now known as Defence Research Agency (Malvern)
Satcom	Satellite communications
SCRA	Single Channel Radio Access
SIU	Subscriber interface unit
SR	Staff Requirement
SSGs	Small Systems Groups
TEACS	Tabuk EUROMUX Area Communications System
TEMPEST	Radiation security. The prevention of unintended radition of information bearing electromagnetic emanations beyond the confibes of an equipment or installation
Tg	Telegraph
TSSP	Tactical Satellite Signal Processor
UKCICC	UK Commanders-in-Chief Committee
ULCE	Unit Line Control Equipment
ULS	Unit Line Switchboard
UNICOM	Unit Computing
UOR	Urgent Operational Requirement
VHF	Very High Frequency
WAN	Wide area network
WMIS	*Wavell* Management Information System
WngOs	Warning Orders

Bibliography

Books

Saddam's War: The Origins of the Kuwait Conflict and the International Response, John Bulloch & Harvey Morris, Faber, 1991, ISBN 0 571 16387 4.

Command and Control for War and Peace, Thomas P. Coakley, National Defense University Press, January 1992, ISBN 0–16–036337–3.

Storm Command, General Sir Peter de la Billière, HarperCollins 1992.

Management Strategies for Information Technology, Michael J. Earl, Prentice-Hall, 1989.

Britain's Gulf War: Operation Granby, Eric Grove [ed.], Harrington Kilbride, 1991.

The Gulf War, Terry Manners, Express Newspapers 1991, ISBN 0–85079–230–4.

Communications and Information Systems for Battlefield Command and Control, M. A. Rice & A. J. Sammes, Brassey's, 1989, ISBN 0–08–036267–2.

Gulf War, Frederick Stanwood, Patrick Allen & Lindsay Peacock, *BCA* 1991.

Command in War, M. Van Creveld, Harvard, 1985.

Military Lessons of the Gulf War, Bruce W. Watson, Bruce George, Peter Tsouras and B.L. Cyr, *BCA* 1991, CN 2533.

The Sunday Times War in the Gulf, a Pictorial History, John Witherow & Aidan Sullivan, Sidgwick & Jackson, 1991, ISBN 0 283 06110 3.

Journal Articles

These have been listed in chronological order of publication under a number of broad category headings.

Lessons Learned

'Critical questions loom in assessing the Gulf War', *Armed Forces Journal International*, 2 March, 1991, **128** (9), pp. 46–7.

'Few flaws in postwar weapons check', *Computerworld*, 18 March, 1991, pp. 1,99.

'Military chiefs seek funding for continued weapons modernizations', *Aerospace Daily*, 20 March, 1991, **157** (55), p. 470.

'War will reshape doctrine, but lessons are limited', *Aviation Week & Space Technology*, 22 April, 1991, pp. 40–3.

'Pentagon weighs key reconnaissance issues highlighted by Gulf War', *Aviation Week & Space Technology*, 22 April, 1991, pp. 78–99.

'Gulf War experience sparks review of RPV priorities', *Aviation Week & Space Technology*, 22 April, 1991, pp 86–7.

'Britain's Gulf role highlights value of flexible tactics, new technology', *Aviation Week & Space Technology*, 22 April, 1991, pp. 104, 107.

'Gulf conflict offered vital insight to success in coalition warfare', *Defense News*, 29 April, 1991, p. 19.

'War stresses need for mobile simulators', *Defense News*, 13 May, 1991, pp. 40, 44.

'The Gulf War: technological and organizational implications', Gene I. Rochlin & Chris C. Demchak, *Survival*, May/June 1991, **33** (3), pp 260–73.

'Key Desert Storm technologies vital to future competitiveness', *Aviation Week & Space Technology*, 3 June 1991, pp. 64–5.

'Lessons from the Gulf – a time for caution', Charles C. Cutshaw, *Jane's Intelligence Review*, July 1991, **3** (7), pp. 314–8.

'Gulf report lists US shortcomings.',John Boatman, *Jane's Defence Weekly*, 27 July 1991, **16** (4), p. 135.

'Operation Desert Storm coalition warfare and lessons learned', *Navy International*, October 1991, **96** (9), pp. 344–8.

Conduct of the Persian Gulf Conflict: An Interim Report to Congress.

C3I and CIS Systems

'C3I warfare moves into new era', *Defense News*, 7 January, 1991.

'Satellites help US forces control the skies over Iraq', *Network World*, 21 January, 1991, pp.1, 46.

'ABC3 system to be last to leave Mid-East', *C4I Report*, 1 April, 1991, **6** (7), pp. 6–7.

'Tactical C3I key to US demolition of Iraq', *C4I Report*, 1 April, 1991, **6** (7), pp. 1–2.

'Electronic warfare played greater role in Desert Storm than any conflict', *Aviation Week & Space Technology*, 22 April, 1991, pp 68–9.

'JSTARS systems were upgraded before deployment', *Defense Daily FT*, 11 February, 1991.

'Six JSTARS ground stations were deployed to Gulf', *Defense Daily FT*, 25 March, 1991.

'Allied forces cracked Iraqi scrambling systems', *Defense News*, 15 April, 1991, p. 19.

'AWACS fleet supplied surveillance data crucial to Allies' Desert Storm victory', *Aviation Week & Space Technology*, 22 April, 1991, p. 82.

' "Filtering" helped top military leaders get proper intelligence information', *Aviation Week & Space Technology*, 22 April, 1991, pp. 84–5.

'Army command, control funds will fall in 1990s', *Defense News*, 22 April, 1991, p. 10.

'Spacecraft played vital role in Gulf War victory', *Aviation Week & Space Technology*, 22 April, 1991, p. 91.

'Command, control advances permeate combat successes', *Signal*, May 1991, **45**, pp. 121–6.

'SPOT images helped allies hit targets in downtown Baghdad', *Armed Forces Journal International*, May 1991, p. 54.

'Military turns to laptop computers to raise fighting effectiveness', *Aviation Week & Space Technology*, 3 June, 1991, pp. 74–5.

' "Granby" and EW options', Martin Streetly, *Jane's Defence Weekly*, 29 June 1991, **15** (26), p. 1178.

'Intergraph in the Gulf', *Computers in Defence*, Autumn 1991, **33**, pp. 22–3.

'Desert navigation devices', *Infantry*, September-October 1991, **81** (5), pp. 37–8.

'Imagination only limit to military, commercial applications for GPS', *Aviation Week & Space Technology*, 14 October 1991, pp. 60–4.

'Desert duty and the digital debate', Sheldon B. Herskovitz, *Journal of Electronic Defense*, November 1991, **14**, pp. 47–54.

'Strategic command and control in the 1990s', D. E. Pearson, *Defense Analysis*, December 1991, **7** (4), pp. 373–99.

'The first space war: the contribution of satellites to the Gulf War', Sir Peter Anson & Dennis Cummings, *RUSI Journal*, Winter 1991, **136**, pp. 45–53.

Scud and Patriot
'Patriot antimissile successes show how software upgrades help meet new threats', *Aviation Week & Space Technology*, 28 January, 1991, pp. 26–8.
'Scud propulsion designs help *Patriot* system succeed', *Aviation Week & Space Technology*, 28 January, 1991, p. 28.
'US Army Patriot proven in new role as anti-tactical ballistic missile weapon', *Aviation Week & Space Technology*, 18 February, 1991, pp. 49–51.
'Iraq's short range surface-to-surface missiles', Duncan Lennox, *Jane's Soviet Intelligence Review*, February 1991, **3**, pp. 58–61.
'Future missiles will outpace *Scuds*', *Defense News*, 4 February, 1991, p. 38.
'Iraq's *Scud* programme – the tip of the iceberg', *Jane's Defence Weekly*, 2 March 1991, **15**, pp. 301–3.
'Iraqi missile operations during "Desert Storm"', Joseph S. Bermudez, *Jane's Soviet Intelligence Review*, March 1991, **3**, pp. 131–5.
'Iraqi missile operations during "Desert Storm" – Update', Joseph S. Bermudez, *Jane's Soviet Intelligence Review*, May 1991, **3**, p. 225.
'Scudbuster: the story behind "Patriot"', Ralph Kinney Bennett, *Reader's Digest*, May 1991, pp. 69–74.
'Success of *Patriot* system shapes debate on future antimissile weapons', *Aviation Week & Space Technology*, 22 April, 1991, pp. 90–1.
'Official says glitch let *Scud* bypass *Patriot*', *Defense News*, 6 May, 1991, p. 20.
'Critics fire misinformation at *Patriot*', *Defense News*, 13 May, 1991, p. 33.

Other Gulf War Weapons
'US fires over 25% of its conventional land attack *Tomahawks* in first week of war', *Aviation Week & Space Technology*, 28 January 1991, pp. 29–30.
'SLAMs hit Iraqi target in first combat firing', *Aviation Week & Space Technology*, 28 January, 1991, pp. 31, 34.
'Technology on trial and a guide to Gulf War weapons', *Flight International*, 13–19 February 1991, pp 30–7.
'Modern bombs in the Gulf', Bill Sweetman, *Jane's Defence Weekly*, 9 February 1991, **15**, p. 178.
'Major weapon systems in the Gulf', *Defense & Armament International*, February/March 1991, (103), pp. 54–9.
'"Steel rain" shut down Iraqi artillery', *Armed Forces Journal International*, May 1991, p. 37.
'Army weapons in "Desert Storm": first assessment', *Defence*, May 1991, **22**, p. 61.

Desert Storm
'Defeating the Iraqis', Wallace Franz, *Armor*, January-February 1991, pp. 8–9.
'Desert Storm: the hundred hours of the Iraqi debacle', *Defense & Armament International*, February/March 1991, (103), pp. 28–32.
'The 100–hour war', Edward M. Flanagan, *Army*, April 1991, **41**, pp. 18, 21–6.
'Desert Storm: a textbook victory', David Eshel, *Military Technology*, April 1991, pp. 28–34.
'The 100–hour war', Dennis Steele, *Army*, April 1991, **41**, pp. 19–20.
'Desert Storm and division air defense', *Armed Forces Journal International*, May 1991, pp. 34, 36.
'Desert Storm: asking the right questions', *Defence*, May 1991, **22**, p. 59.
'Allied strategists altered battle plans to compensate for Dugan's comments', *Aviation Week & Space Technology*, 22 July 1991, p. 60.

'Allies feared massive Iraqi nonconventional attack', *Aviation Week & Space Technology*, 22 July 1991, pp. 61–2.

'Army operations in the Gulf theater', J. J. Yeosock, *Military Review*, September 1991, **71**, pp. 2–15.

'Central command briefing', General H. Norman Schwarzkopf, US Army, Riyadh, Saudi Arabia, Wednesday, 27 February 1991, *Military Review*, September 1991, **71** (9), pp. 65–78.

UK Forces

'1st Armoured Division artillery on Operation Granby', I. G. C. Durie, *Journal of the Royal Artillery*, pp. 16–29.

'UK Forces in the Gulf War', David Miller, *Military Technology*, July 1991, (**15**) 7, pp 39–50.

'Logistics in the Gulf War', Kenneth Hayr, *RUSI Journal*, Autumn 1991, pp. 14–19.

'The principles of multilateral military intervention and the 1990–1991 Gulf crisis', R. M. Connaughton, *British Army Review*, August 1991, (92), pp. 4–10.

'Operation Granby preparation and deployment for war', J. D. Moore-Bick, *Royal Engineers Journal*, December 1991, **105**, pp. 260–7.

Iraqi Forces

'The sword of Saddam, an overview of the Iraqi Armed Forces', John F. Antal, *Armor*, November-December 1990, **99**, pp. 8–12.

'Why Saddam Hussein invaded Kuwait', Efraim Karsh & Inari Rautsi, *Survival*, January/February 1991, **33** (1), pp. 18–30.

'Iraq's mailed fist', John F. Antal, *Infantry*, January-February 1991, **81**, pp. 27–30.

'Iraq's formidable array of guns', *Jane's Defence Weekly*, 2 February 1991, **15**, pp. 136–7.

'Iraq: reasons for the breakdown', *Defense & Armament International*, No 103, February/March 1991, (103), pp. 20–1.

'Tactical defensive doctrine of the Iraqi ground forces', Richard Philips, *Jane's Soviet Intelligence Review*, March 1991, **3**, pp. 116–9.

'Think twice about trying Saddam', *Armed Forces Journal International*, April 1991, p. 28.

'Iraqi intelligence and security services', Andrew Rathmell, *International Defense Review*, May 1991, **24** (5), pp. 393–5.

'The Iraqi Army's defeat in Kuwait', James W. Pardew, *Parameters*, Winter 1991–92, **21**, pp. 17–23.

The Air War

'Why the air war worked', *Armed Forces Journal International*, April 1991, pp. 44–5.

'Flexibility of attack aircraft crucial to crushing Iraq's military machine', *Aviation Week & Space Technology*, 22 April 1991, pp. 46–7.

'F-117A pilots conduct precision bombing in high threat environment', *Aviation Week & Space Technology*, 22 April 1991, pp. 51, 53.

'Air crew training, avionics credited for F-15E's high target hit rates', *Aviation Week & Space Technology*, 22 April 1991, pp. 54–5.

'The Gulf air campaign – an overview', Niall Irving, *RUSI Journal*, February 1992, **137**, pp. 10–14.

Background and Chronology

'Chronology', *Military Review*, September 1991, **71** (9), pp. 65–78.

UN Resolutions, *Military Review*, September 1991, **71** (9), p. 79.

Forces Committed, *Military Review*, September 1991, **71** (9), pp. 80–1.

1 (BR) Corps

Index